THE CLINICAL MEASUREMENT PACKAGE
A Field Manual

The Dorsey Series in Social Welfare

THE Clinical Measurement Package
A Field Manual

WALTER W. HUDSON, Ph.D.
School of Social Work
Florida State University

1982
THE DORSEY PRESS
Homewood, Illinois 60430

TO MYRNA, KIRK, AND KAREN
The most special people in my life
(who also know more about their
depression and self-esteem than
they ever wanted to know!)

Library of Congress Catalog Card No. 81-70422
Printed in the United States of America

1 2 3 4 5 6 7 8 9 0 D 9 8 7 6 5 4 3 2

Preface

This manual presents and describes nine short-form scales that were designed for repeated use with a client to monitor and evaluate progress in therapy. The scales in this manual were designed to measure the severity or magnitude of problems that clients have with (1) depression, (2) self-esteem, (3) marital discord, and (4) sexual discord; (5) parent-child relationships as seen by the parent, (6) as seen by the child in relation to the mother, (7) and as seen by the child in relation to the father; (8) intrafamilial stress; and (9) peer relationships. The nine scales are exhibited at the end of Chapter 1 and Chinese, French, German, and Spanish translations of the scales are available from the publisher.*

The development and validation work for this collection of scales was begun during the mid-1970s, and the first validation study was published two years later (Hudson and Glisson, 1976). In the same year, another study was released (Hudson and Proctor, 1976a), and it was briefly described in a separate publication (Hudson and Proctor, 1977). By early 1977, a total of seven scales had been developed and partially validated, and those seven scales were described in a paper, "A Measurement Package for Clinical Workers," that was presented at the Council on Social Work Education Annual Program Meeting held in Phoenix, Arizona (Hudson, 1977). By the middle of the next year, two more scales had been developed, and the above paper was revised to include all nine of the scales that were collectively refer-

*The Dorsey Press, 1818 Ridge Road, Homewood, Illinois 60430.

red to as the *Clinical Measurement Package* or CMP scales (Hudson, 1978a, 1981a).

Since the initial publication and release of the CMP scales, they have been favorably received, and I have received numerous requests for information about their availability, use, reliability, and validity. During the three-year period following the presentation of the first seven scales in 1977, I personally mailed or distributed nearly 2,000 copies of the first CMP papers to individuals who directly inquired about them or who sought to use them in the classroom, in agencies, or in private practice. Thus far, it appears that the CMP scales are currently being used by social workers, psychologists, psychiatrists, pastoral counselors, and other helping professionals in public and private social welfare agencies, schools, hospitals, medical clinics, correctional institutions, and private practice across the United States and in more than 15 foreign countries. In addition, the CMP scales have been reproduced in several practice and research texts that are currently being used for classroom teaching purposes (Fischer, 1978; Grinnell, 1981; Rosenblatt and Waldfogel, 1982; Fischer and Bloom, 1982).

During the past seven years, much work has been devoted to the validation of the CMP scales, and while it is doubtful that any measurement scale, such as those found in this manual, will ever be *completely* validated, enough data have been obtained to strongly support the validity and reliability of the entire package. Moreover, in the process of conducting several independent studies aimed at the validation of the CMP scales, much information has been obtained concerning their use and performance. Thus, this field manual has been prepared as a means to further describe the CMP scales, provide initial information to those who are not familiar with them and give up-to-date information to those who have used or considered the scales previously.

This field manual was designed to serve as a reference to be consulted when using the CMP scales in clinical applications and when planning, designing, or conducting research. However, it can also be used as a supplemental text for classroom and practicum training. When used as a supplemental text in practice and in research courses, it has been found that the material in this manual is highly effective in providing students with a distinct technology that facilitates their learning of the major content of the course. It also provides them with immediate, easily mastered beginning knowledge of procedures they can use to determine whether their clients are making progress in treatment. It does that largely by demonstrating how they can obtain and use feedback information as a basis for modifying treatment to enhance the likelihood of a more positive outcome for the client. When these materials are presented in a first or second semester practice course, it is usually found that students make good use of this technology throughout the remainder of their professional training and in their professional practice following the completion of training.

In addition to the use as a supplementary text in programs of classroom and practicum training, this manual can also be used effectively as a training aid for consultants and continuing education instructors who work with formal organizations in a variety of human service fields. The CMP scales and the introductory materials contained in Chapters 1 through 5 provide a convenient way of introducing the use of formal measurement procedures into a variety of professional practice modalities for those professional personnel who have not been previously trained in the use of such technology. Professional therapists who already know how to use single-subject designs and formal measurement tools often find the scales to be of interest and use in their work.

It should be observed that this manual was organized primarily for use by clinicians. Chapters 1 through 5 present the most important information that therapists need in order to administer, score, interpret, and use the CMP scales in clinical practice. Chapters 6 through 8, it is hoped, will be of interest to therapists because they need to know about and understand the technical psychometric properties and performance of the scales. However, these three chapters are oriented more for use by consultants and researchers who need not only a report of findings but a description of how they were obtained and the logic or rationale behind the methods that were used. Although these three chapters may be irritatingly technical for those who wish only to review the final results, an effort has been made to present the material in a partially instructional format for those who are not trained in research technology but who want to know more about this part of the work.

The proper use of the CMP scales and the material in this manual is easily understood and mastered by qualified professional therapists. However, these measurement tools are not intended for use by untrained individuals. The scales described in this manual are simple but very powerful devices that are capable of revealing both minor and serious problems that people have in several areas of personal and social functioning. They are not intended for use by persons who are not trained to deal with such problems, and they should be used only by competent professionals, researchers, scholars, and those who are engaged in supervised study and training.

WALTER W. HUDSON

Acknowledgments

In undertaking the work that goes into such a long project, it is amazing how indebted one becomes to others. First among them are my colleagues who actually participated in and conducted the validation research with me: Dianne F. Harrison, Enola K. Proctor, Charles A. Giuli, Paul P. L. Cheung, Barbara Wung, Marianne Borges, Paul C. Crosscup, Joseph D. Acklin, Jennie C. Bartosh, Roger Hamada, Ruth Keech, Jon Harlan, Neil Abell, Barry Jones, and Paula S. Nurius. While each is clearly acknowledged as a coauthor of the various studies that are referred to in this manual, they have my very special thanks and gratitude for the hard work performed and their superb collegial support and interest. Paul P. L. Cheung, Jean-Marie Boisvert, Ralf Lutgehetmann, and Rufina Ada Montemayor provided, respectively, the Chinese, French, German, and Spanish language translations of the CMP scales, and their special assistance in this work is gratefully acknowledged.

Although a large number of people have supported and encouraged the development and validation of the CMP scales, a note of thanks is extended to Joel Fischer who served as a friend and critic in promoting the use and testing of the CMP scales in a wide variety of research and clinical settings. I particularly wish to thank Shanti K. Khinduka, Daniel S. Sanders, and Donald R. Bardill who provided the institutional supports that were needed for the conduct of this work. I am especially grateful to Beulah Compton, Dean Hepworth, Joy Johnson, Paula S. Nurius, and Charles Zastrow who, on several occasions, prevented me from making worse mistakes than those that may still remain. Finally, I thank the large number of colleagues and students across the country who have expressed interest and encouragement for this work, who used and tested the scales in many different applications and settings, and who provided assessment, feedback, and suggestions that were helpful in one way or another. I wish they all might be named.

W.W.H.

Contents

1 Introduction

This field manual provides information about the use and performance of nine shortform measurement scales that were designed for use in assessing the severity or magnitude of a variety of personal and social problems. The nine scales are collectively referred to as the *Clinical Measurement Package* or CMP. Each scale was designed for use by therapists and researchers who are in need of reliable and valid measures of several important variables that define and influence the quality of personal and social functioning among individuals, couples, families, and small groups.

Each of the CMP scales is a paper-pencil, self-report questionnaire that was designed to measure the degree, severity, or magnitude of a distinct and separate problem in personal and social functioning. Each scale has the same format and structure, and each one has 25 items. This particular length was selected for three specific reasons: the scales are long enough to produce good reliabilities; they are short enough to be used in repeated administrations with the same client on a regular or periodic basis; and the use of 25 items leads to the construction of a very simple scoring procedure. Each scale is scored in exactly the same manner. Only one scoring formula is needed for all nine scales, and each scale is scored to have a score range from 0 to 100 where a low score indicates the relative absence of the problem being measured, and higher scores indicate the presence of a more severe problem. Each scale uses essentially the same set of instructions to improve ease of administration and to promote uniformity of client responses.

Another desirable characteristic of the CMP scales is that each one has a clinical cutting score of 30. That is, it is generally found that persons who obtain a score above 30 have a clinically significant problem in the area being measured, while those who score below 30 are generally free of such problems. The existence of such a clinical cutting score is important to both clinicians and researchers. It provides both a diagnostic benchmark and a criterion against which to judge the effectiveness of treatment.

Table 1–1 presents the name of each scale, a three-character label for each scale, and the construct it was designed to measure. This table was prepared as a quick reference to the entire CMP, and each of the scales is presented at the end of this chapter.

In developing the CMP scales, it was recognized that any measurement tool that is used to characterize human problems or to make decisions about them must have at least two major psychometric characteristics: it must be reliable, and it must be valid. However, if a scale is to be used repeatedly with the same client to monitor and evaluate the severity of the client's problem, it must have a number of other desirable characteristics (Hudson, 1981b): it must be short; it must be easy to administer; it must be easy to score; it must be easy to understand and interpret; and it must not suffer response decay when used repeatedly over many occasions. Each of the CMP scales meets all of these psychometric requirements. Each scale has a reliability of .90 or better, and they all have good content, concurrent, factorial, discriminant, and construct validity.

There is one exception to the above general claims concerning the reliability and validity of the CMP scales. As of this date, there are no available data concerning the psychometric characteristics of the *Index of Peer Relations* (IPR) scale. A study is currently in progress

TABLE 1–1
The CMP scales

Label	Scale name	Measurement construct
GCS	Generalized Contentment Scale	Depression
ISE	Index of Self-Esteem	Self-esteem problem
IMS	Index of Marital Satisfaction	Marital discord
ISS	Index of Sexual Satisfaction	Sexual discord
IPA	Index of Parental Attitudes	Discord with child
CAM	Child's Attitude toward Mother	Discord with mother
CAF	Child's Attitude toward Father	Discord with father
IFR	Index of Family Relations	Intrafamilial stress
IPR	Index of Peer Relations	Peer discord

to obtain that information (Nurius and Hudson, 1982), and it is expected that a final research report will be available within a year of the publication of this manual. The IPR has been included as part of the CMP because it is very likely that it will have the same general psychometric characteristics as the other eight CMP scales.

Although the CMP scales can be used in a wide variety of situations and settings, the initial purpose for their development was for use in single-subject, repeated measures designs to monitor and guide the course of treatment with individuals, couples, families, and small groups. While this was the primary purpose for the development of the CMP scales, they can also be used in the conduct of planned group experiments, surveys, and comparative studies. The scales appear to be especially useful in clinical settings as aids to developing and confirming diagnosis, to evaluate change, and as concomitant assessment tools when the primary focus of treatment is not on one or more of the dysfunctions or problems measured by the CMP. The next sections present a brief description of each scale.

NONPSYCHOTIC DEPRESSION

The *Generalized Contentment Scale* or GCS was designed to measure the degree, severity, or magnitude of nonpsychotic depression. This does not mean the scale cannot be used with a client who has been diagnosed as having psychosis. Rather, the scale should not be used during the time a client is actively psychotic. The GCS can be used with any client following remission of a psychotic episode or crisis, but it should not be used with clients who are chronically psychotic. Clinical experience to date indicates that therapists should be concerned about the possibility of suicide or a suicidal attempt for clients who score above 70 on the GCS. Suicidal ideation is often found among clients whose GCS scores exceed 50. These statements should not be taken to mean that the GCS predicts suicidal attempts. Rather, it has been found that for a number of persons who did attempt suicide, the GCS score exceeded 70 immediately following the suicidal episode.

SELF-ESTEEM PROBLEMS

The *Index of Self-Esteem* or ISE was designed to measure the degree, severity, or magnitude of a problem the client has with self-esteem. Thus, a high score on the ISE indicates the client has a self-esteem problem; a low sense of self-esteem. In using the ISE scale, it is important to make a distinction between self-concept and self-esteem. Self-esteem, as conceptualized and measured with respect to the ISE, is the evaluative component of self-concept. The client may have a very accurate self-concept and a severe problem with self-esteem. For example, a client may see himself accurately as a domineering, demanding husband (self-concept); if he likes and admires domineering and demanding people, he is likely to have a high sense of self-esteem. However, if he dislikes such people, he is likely to have a low sense of self-esteem. Because the ISE and GCS scales tend to correlate highly with one another, the GCS scale should be administered to clients who have very large ISE scores in order to evaluate the severity of a probable depressive reaction. However, these scales measure two entirely different problems and must not be used as substitutes for one another.

MARITAL DISCORD

The *Index of Marital Satisfaction* or IMS was designed to measure the degree, severity, or magnitude of a problem a spouse or partner has in the marital relationship. The IMS does not measure or characterize the dyadic relationship as a single entity but measures the magnitude of marital discord or dissatisfaction that is felt or perceived by one partner. It is therefore quite possible, indeed quite common, to find that one partner will have a very high IMS score while the other partner will have a much lower score. It is difficult to make an empirically valid distinction between the concepts of marital dissatisfaction and marital discord, and for purposes of evaluating or interpreting the IMS score these two concepts are equated. The IMS must not be treated as a marital adjustment measure. A couple may have a good marital adjustment in the sense that they have arrived at some satisfactory arrangement for living and working together. Such couples may nonetheless display a high degree of marital discord or dissatisfaction.

SEXUAL DISCORD

The *Index of Sexual Satisfaction* or ISS was designed to measure the degree, severity, or magnitude of a problem in the sexual component of a dyadic relationship. It measures the degree of sexual discord or dissatisfaction that is felt or perceived by a client with respect to the sexual relationship with a spouse or partner. The items on the ISS were designed so that they will not be offensive or repugnant to people having very different moral convictions and attitudes with respect to human sexual behavior. Although the IMS and ISS scales tend to be highly correlated with one another, they measure two entirely different problems, and neither of these scales can be used as a substitute for the other.

PARENT-CHILD RELATIONSHIP PROBLEMS

There are three separate scales that were designed to measure the degree, severity, or magnitude of a problem in a parent-child relationship. The *Index of Parental Attitudes* or IPA is completed by a parent with respect to the parent's relationship with a specific child. It measures the degree, severity, or magnitude of a problem the parent has in the relationship with a child—regardless of the age of the child. The child may be an infant, young child, adolescent, or an adult.

The *Child's Attitude toward Mother* or CAM and the *Child's Attitude toward Father* or CAF scales were designed to measure the degree, severity, or magnitude of a problem a child has with its mother or father. At this time, there are few scales that are designed to treat the child as a primary source of information about the severity of parent-child relationship problems, and the CAM and CAF scales were designed partially to fill that vacuum. The CAM and CAF scales can be used with adolescent and adult children, but it is recommended they never be used with children under the age of 12.

FAMILY RELATIONSHIP PROBLEMS

The *Index of Family Relations* or IFR was designed to measure the degree, severity, or magnitude of a problem that family members have in their relationships with one another as felt or perceived by the client. This scale permits a client to characterize the severity of family relationship problems in a global fashion and can be regarded as a measure of intrafamilial stress. It can be used productively with one client, or it can be used in conjoint therapy with two or more family members. The IFR can be used as a measure of the familial environment of the client (a rough index of the quality of family life for, and as perceived by, the client), and it can be used in helping the client to deal with problems in relating to the family as a whole.

PEER RELATIONSHIP PROBLEMS

The *Index of Peer Relations* or IPR was designed to measure the degree, severity, or magnitude of a problem the client has in relationships with peers. It can be used as a global measure of peer relationship problems or the therapist can specify the peer reference group. For example, the client might be asked to complete the IPR in terms of the way the client feels about work associates—or friends, classmates, et al. If a peer reference group is specified by the therapist, a note to that effect should be placed at the top of the IPR in the space provided. The IPR can be used to measure a client's peer relationship problems with respect to more than one reference group. For example, the client might be asked to complete the IPR once in relation to a friendship group, again in relation to a work group, and yet again in relation to a leisure activity group.

RESTRICTIONS

There are few formal restrictions to be placed on the use of any of the CMP scales but some must always be taken into account. The first restriction is that none of the CMP scales should be completed by persons under the age of 12 years. The literacy skills, cognitive development, and ability to integrate affective responses with the item content and meaning of each of the scales demand a level of maturity not usually found in persons under 12 years of age. If one is working with an unusually precocious child under the age of 12 and is convinced that the person's responses would be valid, then by all means use any of the CMP scales deemed appropriate. However, a decision to use the scales with such individuals should be made only after the therapist has acquired extensive experience in using them with older persons.

A second restriction is that none of these scales should be used with mentally retarded persons. Mentally retarded individuals rarely give valid responses to the CMP scales even though such persons may function very well in the world of work and may do a good job of maintaining and taking care of themselves. This does not, of course, exclude their use with brain damaged individuals who have good affective and cognitive functioning.

A third restriction is that the CMP scales should be used with great caution in situations wherein the therapist is in a position to give or withhold important or valued social sanctions and the client's responses to the scales can be seen as having any possible influence on the therapist's decisions in that regard. For example, the social worker, psychologist, or therapist who must make some recommendation concerning parole of an inmate or the release of a legally committed mental patient may elicit wholly invalid responses if a favorable decision is to be based on symptom remission and the respondent is seeking escape or release from confinement. Under such circumstances, the validity of the CMP scale responses must be carefully judged against any and all other evidence concerning the client's progress in treatment.

OTHER LANGUAGE VERSIONS

Over the past few years, the CMP scales have been used in clinical and research applications with individuals, couples, families, and small groups whose primary language is not English. Because it appears that these scales are highly useful in working with non-English speaking populations, each of them has been translated into several other languages: Chinese, French, German, and Spanish. Paul P. L. Cheung at the Center for Population Studies at the University of Michigan, Ann Arbor, Michigan, prepared the Chinese translations; Jean-Marie Boisvert at the Hôpital Louis-H. LaFontaine, Montreal, Quebec, Canada prepared the French translations; Ralf Lutgehetmann at the Psychologisches Institut, der Universitat Heidelberg, Heidelberg, Germany, prepared the German translations, and Rufina Ada Montemayor at the M. D. Anderson Hospital and Tumor Institute, Houston, Texas, prepared the Spanish translations.

ACQUISITION AND USE OF THE SCALES

Each of the CMP scales is copyrighted and reproduced by The Dorsey Press. Those who would like to obtain a trial set of the CMP scales can do so by writing to The Dorsey Press, 1818 Ridge Road, Homewood, Illinois 60430.

GENERALIZED CONTENTMENT SCALE (GCS) Today's Date ________

NAME ______________________________

This questionnaire is designed to measure the degree of contentment that you feel about your life and surroundings. It is not a test, so there are no right or wrong answers. Answer each item as carefully and accurately as you can by placing a number beside each one as follows:

1 Rarely or none of the time
2 A little of the time
3 Some of the time
4 A good part of the time
5 Most or all of the time

Please begin.

1. I feel powerless to do anything about my life. ____
2. I feel blue. ____
3. I am restless and can't keep still. ____
4. I have crying spells. ____
5. It is easy for me to relax. ____
6. I have a hard time getting started on things that I need to do. ____
7. I do not sleep well at night. ____
8. When things get tough, I feel there is always someone I can turn to. ____
9. I feel that the future looks bright for me. ____
10. I feel downhearted. ____
11. I feel that I am needed. ____
12. I feel that I am appreciated by others. ____
13. I enjoy being active and busy. ____
14. I feel that others would be better off without me. ____
15. I enjoy being with other people. ____
16. I feel it is easy for me to make decisions. ____
17. I feel downtrodden. ____
18. I am irritable. ____
19. I get upset easily. ____
20. It is hard for me to have a good time. ____
21. I have a full life. ____
22. I feel that people really care about me. ____
23. I have a great deal of fun. ____
24. I feel great in the morning. ____
25. I feel that my situation is hopeless. ____

5, 8, 9, 11, 12, 13, 15, 16, 21, 22, 23, 24

unidemensional scale.

INDEX OF SELF-ESTEEM (ISE)

Today's Date ________

NAME ________________________________

This questionnaire is designed to measure how you see yourself. It is not a test, so there are no right or wrong answers. Please answer each item as carefully and accurately as you can by placing a number by each one as follows:

1 Rarely or none of the time
2 A little of the time
3 Some of the time
4 A good part of the time
5 Most or all of the time

Please begin.

1. I feel that people would not like me if they really knew me well. ____
2. I feel that others get along much better than I do. ____
3. I feel that I am a beautiful person. ____
4. When I am with other people I feel they are glad I am with them. ____
5. I feel that people really like to talk with me. ____
6. I feel that I am a very competent person. ____
7. I think I make a good impression on others. ____
8. I feel that I need more self-confidence. ____
9. When I am with strangers I am very nervous. ____
10. I think that I am a dull person. ____
11. I feel ugly. ____
12. I feel that others have more fun than I do. ____
13. I feel that I bore people. ____
14. I think my friends find me interesting. ____
15. I think I have a good sense of humor. ____
16. I feel very self-conscious when I am with strangers. ____
17. I feel that if I could be more like other people I would have it made. ____
18. I feel that people have a good time when they are with me. ____
19. I feel like a wallflower when I go out. ____
20. I feel I get pushed around more than others. ____
21. I think I am a rather nice person. ____
22 I feel that people really like me very much. ____
23. I feel that I am a likeable person. ____
24. I am afraid I will appear foolish to others. ____
25. My friends think very highly of me. ____

3, 4, 5, 6, 7, 14, 15, 18, 21, 22, 23, 25

INDEX OF MARITAL SATISFACTION (IMS) Today's Date ________

NAME ______________________________

This questionnaire is designed to measure the degree of satisfaction you have with your present marriage. It is not a test, so there are no right or wrong answers. Answer each item as carefully and as accurately as you can by placing a number beside each one as follows:

1 Rarely or none of the time
2 A little of the time
3 Some of the time
4 A good part of the time
5 Most or all of the time

Please begin.

1. I feel that my partner is affectionate enough. ____
2. I feel that my partner treats me badly. ____
3. I feel that my partner really cares for me. ____
4. I feel that I would not choose the same partner if I had it to do over again. ____
5. I feel that I can trust my partner. ____
6. I feel that our relationship is breaking up. ____
7. I feel that my partner doesn't understand me. ____
8. I feel that our relationship is a good one. ____
9. I feel that ours is a very happy relationship. ____
10. I feel that our life together is dull. ____
11. I feel that we have a lot of fun together. ____
12. I feel that my partner doesn't confide in me. ____
13. I feel that ours is a very close relationship. ____
14. I feel that I cannot rely on my partner. ____
15. I feel that we do not have enough interests in common. ____
16. I feel that we manage arguments and disagreements very well. ____
17. I feel that we do a good job of managing our finances. ____
18. I feel that I should never have married my partner. ____
19. I feel that my partner and I get along very well together. ____
20. I feel that our relationship is very stable. ____
21. I feel that my partner is a comfort to me. ____
22. I feel that I no longer care for my partner. ____
23. I feel that the future looks bright for our relationship. ____
24. I feel that our relationship is empty. ____
25. I feel there is no excitement in our relationship. ____

1, 3, 5, 8, 9, 11, 13, 16, 17, 19, 20, 21, 23

INDEX OF SEXUAL SATISFACTION (ISS) Today's Date ________

NAME ______________________________

This questionnaire is designed to measure the degree of satisfaction you have in the sexual relationship with your partner. It is not a test, so there are no right or wrong answers. Answer each item as carefully and accurately as you can by placing a number beside each one as follows:

1 Rarely or none of the time
2 A little of the time
3 Some of the time
4 A good part of the time
5 Most or all of the time

Please begin.

1. I feel that my partner enjoys our sex life. ____
2. My sex life is very exciting. ____
3. Sex is fun for my partner and me. ____
4. Sex with my partner has become a chore for me. ____
5. I feel that sex is dirty and disgusting. ____
6. My sex life is monotonous. ____
7. When we have sex it is too rushed and hurriedly completed. ____
8. I feel that my sex life is lacking in quality. ____
9. My partner is sexually very exciting. ____
10. I enjoy the sex techniques that my partner likes or uses. ____
11. I feel that my partner wants too much sex from me. ____
12. I think that sex is wonderful. ____
13. My partner dwells on sex too much. ____
14. I try to avoid sexual contact with my partner. ____
15. My partner is too rough or brutal when we have sex. ____
16. My partner is a wonderful sex mate. ____
17. I feel that sex is a normal function of our relationship. ____
18. My partner does not want sex when I do. ____
19. I feel that our sex life really adds a lot to our relationship. ____
20. My partner seems to avoid sexual contact with me. ____
21. It is easy for me to get sexually excited by my partner. ____
22. I feel that my partner is sexually pleased with me. ____
23. My partner is very sensitive to my sexual needs and desires. ____
24. My partner does not satisfy me sexually. ____
25. I feel that my sex life is boring. ____

1, 2, 3, 9, 10, 12, 16, 17, 19, 21, 22, 23

INDEX OF PARENTAL ATTITUDES (IPA) Today's Date ________

PARENT'S NAME ______________________ CHILD'S NAME ______________________

This questionnaire is designed to measure the degree of contentment you have in your relationship with your child. It is not a test, so there are no right or wrong answers. Answer each item as carefully and accurately as you can by placing a number beside each one as follows:

1 Rarely or none of the time
2 A little of the time
3 Some of the time
4 A good part of the time
5 Most or all of the time

Please begin.

1. My child gets on my nerves. ____
2. I get along well with my child. ____
3. I feel that I can really trust my child. ____
4. I dislike my child. ____
5. My child is well behaved. ____
6. My child is too demanding. ____
7. I wish I did not have this child. ____
8. I really enjoy my child. ____
9. I have a hard time controlling my child. ____
10. My child interferes with my activities. ____
11. I resent my child. ____
12. I think my child is terrific. ____
13. I hate my child. ____
14. I am very patient with my child. ____
15. I really like my child. ____
16. I like being with my child. ____
17. I feel like I do not love my child. ____
18. My child is irritating. ____
19. I feel very angry toward my child. ____
20. I feel violent toward my child. ____
21. I feel very proud of my child. ____
22. I wish my child was more like others I know. ____
23. I just do not understand my child. ____
24. My child is a real joy to me. ____
25. I feel ashamed of my child. ____

2, 3, 5, 8, 12, 14, 15, 16, 21, 24

CHILD'S ATTITUDE TOWARD MOTHER (CAM) Today's Date ________

NAME ______________________________

This questionnaire is designed to measure the degree of contentment you have in your relationship with your mother. It is not a test, so there are no right or wrong answers. Answer each item as carefully and accurately as you can by placing a number beside each one as follows:

1 Rarely or none of the time
2 A little of the time
3 Some of the time
4 A good part of the time
5 Most or all of the time

Please begin.

1. My mother gets on my nerves. ____
2. I get along well with my mother. ____
3. I feel that I can really trust my mother. ____
4. I dislike my mother. ____
5. My mother's behavior embarrasses me. ____
6. My mother is too demanding. ____
7. I wish I had a different mother. ____
8. I really enjoy my mother. ____
9. My mother puts too many limits on me. ____
10. My mother interferes with my activities. ____
11. I resent my mother. ____
12. I think my mother is terrific. ____
13. I hate my mother. ____
14. My mother is very patient with me. ____
15. I really like my mother. ____
16. I like being with my mother. ____
17. I feel like I do not love my mother. ____
18. My mother is very irritating. ____
19. I feel very angry toward my mother. ____
20. I feel violent toward my mother. ____
21. I feel proud of my mother. ____
22. I wish my mother was more like others I know. ____
23. My mother does not understand me. ____
24. I can really depend on my mother. ____
25. I feel ashamed of my mother. ____

2, 3, 8, 12, 14, 15, 16, 21, 24

CHILD'S ATTITUDE TOWARD FATHER (CAF) Today's Date ________

NAME ______________________________

This questionnaire is designed to measure the degree of contentment you have in your relationship with your father. It is not a test, so there are no right or wrong answers. Answer each item as carefully and accurately as you can by placing a number beside each one as follows:

1 Rarely or none of the time
2 A little of the time
3 Some of the time
4 A good part of the time
5 Most or all of the time

Please begin.

1. My father gets on my nerves. ____
2. I get along well with my father. ____
3. I feel that I can really trust my father. ____
4. I dislike my father. ____
5. My father's behavior embarrasses me. ____
6. My father is too demanding. ____
7. I wish I had a different father. ____
8. I really enjoy my father. ____
9. My father puts too many limits on me. ____
10. My father interferes with my activities. ____
11. I resent my father. ____
12. I think my father is terrific. ____
13. I hate my father. ____
14. My father is very patient with me. ____
15. I really like my father. ____
16. I like being with my father. ____
17. I feel like I do not love my father. ____
18. My father is very irritating. ____
19. I feel very angry toward my father. ____
20. I feel violent toward my father. ____
21. I feel proud of my father. ____
22. I wish my father was more like others I know. ____
23. My father does not understand me. ____
24. I can really depend on my father. ____
25. I feel ashamed of my father. ____

2, 3, 8, 12, 14, 15, 16, 21, 24

INDEX OF FAMILY RELATIONS (IFR) Today's Date ________

NAME ________________________________

This questionnaire is designed to measure the way you feel about your family as a whole. It is not a test, so there are no right or wrong answers. Answer each item as carefully and accurately as you can by placing a number before each one as follows:

1 Rarely or none of the time
2 A little of the time
3 Some of the time
4 A good part of the time
5 Most or all of the time

Please begin.

1. The members of my family really care about each other. ____
2. I think my family is terrific. ____
3. My family gets on my nerves. ____
4. I really enjoy my family. ____
5. I can really depend on my family. ____
6. I really do not care to be around my family. ____
7. I wish I was not part of this family. ____
8. I get along well with my family. ____
9. Members of my family argue too much. ____
10. There is no sense of closeness in my family. ____
11. I feel like a stranger in my family. ____
12. My family does not understand me. ____
13. There is too much hatred in my family. ____
14. Members of my family are really good to one another. ____
15. My family is well respected by those who know us. ____
16. There seems to be a lot of friction in my family. ____
17. There is a lot of love in my family. ____
18. Members of my family get along well together. ____
19. Life in my family is generally unpleasant. ____
20. My family is a great joy to me. ____
21. I feel proud of my family. ____
22. Other families seem to get along better than ours. ____
23. My family is a real source of comfort to me. ____
24. I feel left out of my family. ____
25. My family is an unhappy one. ____

1, 2, 4, 5, 8, 14, 15, 17, 18, 20, 21, 23

INDEX OF PEER RELATIONS (IPR) Today's Date ________

NAME ______________________________ GROUP ______________________

This questionnaire is designed to measure the way you feel about the people you work, play, or associate with most of the time; your peer group. It is not a test so there are no right or wrong answers. Answer each item as carefully and as accurately as you can be placing a number beside each one as follows:

1 Rarely or none of the time
2 A little of the time
3 Some of the time
4 A good part of the time
5 Most or all of the time

Please begin.

1. I get along very well with my peers. ____
2. My peers act like they don't care about me. ____
3. My peers treat me badly. ____
4. My peers really seem to respect me. ____
5. I don't feel like I am "part of the group". ____
6. My peers are a bunch of snobs. ____
7. My peers understand me. ____
8. My peers seem to like me very much. ____
9. I really feel "left out" of my peer group. ____
10. I hate my present peer group. ____
11. My peers seem to like having me around. ____
12. I really like my present peer group. ____
13. I really feel like I am disliked by my peers. ____
14. I wish I had a different peer group. ____
15. My peers are very nice to me. ____
16. My peers seem to look up to me. ____
17. My peers think I am important to them. ____
18. My peers are a real source of pleasure to me. ____
19. My peers don't seem to even notice me. ____
20. I wish I were not part of this peer group. ____
21. My peers regard my ideas and opinions very highly. ____
22. I feel like I am an important member of my peer group. ____
23. I can't stand to be around my peer group. ____
24. My peers seem to look down on me. ____
25. My peers really do not interest me. ____

1, 4, 7, 8, 11, 12, 15, 16, 17, 18, 21, 22

2 Scoring and interpreting the CMP scales

In this chapter procedures are set forth for scoring and interpreting each of the CMP scales. The psychometric structure of the scales is also described. Further details concerning their use and interpretation are presented in later chapters.

STRUCTURE AND SCORING

Each of the CMP scales is structured as a 25-item summated category partition scale (Stevens, 1968) wherein each item is scored according to the following five categories: 1 = rarely or none of the time; 2 = a little of the time; 3 = some of the time; 4 = a good part of the time; and 5 = most or all of the time. For each scale, some of the items are positively worded statements or descriptors, and others are negatively worded to partially control for response set biases. All of the items were randomly ordered within each scale, and it should be noted that the CAM and CAF scales are identical—except for the substitution of the word *mother* or *father* on the appropriate scale.

Although each of the CMP scales was constructed as a multi-item scale, none was designed as multidimensional measures. Each was designed to provide only a single-dimension characterization of the degree or magnitude of a personal or social problem, and a detailed account of the uni- and multidimensional characteristics of the CMP scales is provided later.

Since each of the CMP scales is a unidimensional measure of a personal or social problem and since each was designed to measure the degree, severity, or magnitude of such problems, the CMP scales can be seen to function much like a thermometer. They will provide information about the degree or magnitude of a problem but they will not provide any information about the cause, source, type, or origin of a problem. Some may believe this limitation is a serious one, but so be it. A thermometer provides no information about the source or cause of the heat that it measures. Yet, thermometers are very important and useful measurement tools.

Because each of the CMP scales was designed to measure the degree or magnitude of a personal or social problem, each scale is scored so that higher scores represent more severe problems, and lower scores indicate the relative absence of such problems. The first step in scoring any of the CMP scales is to reverse-score each of the positively worded items so that an item score of 5 becomes 1, 4 becomes 2, 2 becomes 4, 1 becomes 5, and a score of 3 remains unchanged. A simple procedure for reverse-scoring the appropriate items is to subtract the item score from 6. That is, if X is the original score for an item and Y is its score after reversal, the reversal can be computed as

$$Y = 6 - X. \tag{2.1}$$

Because the items that must be reverse-scored differ from one scale to the next, the numbers of those items have been listed below the copyright notation at the bottom of each scale.

After reverse-scoring the appropriate items on any scale and then denoting the item responses as Y, the total score, S, is computed for each scale as

$$S = \Sigma Y - 25. \tag{2.2}$$

This very simple scoring formula has the advantage of producing a range of values from 0 to 100. The scoring formula shown as equation 2.2 should be used *only* for clients who complete every item on one of the CMP scales. The overwhelming majority of all clients complete all 25 items, so in nearly all cases it is possible to use this simple scoring procedure.

If a client fails to complete one or more items on any of the CMP scales, it is necessary to use a different scoring procedure. After reverse-scoring the appropriate items on any scale and again denoting the item responses as Y, the total score, S, is computed as

$$S = (\Sigma Y - N)(100)/[(N)(4)] \tag{2.3}$$

where N is the number of items that were properly completed by the respondent; any item that is left blank or scored outside the range from 1 to 5 is given a score of zero (0).

There are two distinct advantages in using the scoring formula shown as equation 2.3. First, it produces for each CMP scale a score that has a range from 0 to 100. This feature facilitates both the comparison and interpretation of scores obtained from different scales. Second, use of this scoring formula will always produce a score range of 0 to 100 regardless of the number of items that a respondent omits or fails to properly complete. The effect of the scoring formula shown as equation 2.3 is to replace omitted or improperly scored items with the mean of the properly completed items.

Both research and clinical experience with the CMP scales over the years indicates that if a client agrees to complete one or more of the scales, it is exceedingly rare that the client will omit more than five items. Moreover, the omission of five or fewer items on any of the scales has a negligible effect on the scales' reliability and validity. This feature of the scales arises from the fact that they all have good content validity and they are constructed on the basis of the domain sampling theory of measurement (Nunnally, 1978); the omission of a few items does not destroy the ability of a scale to represent the domain from which its item content was drawn. However, if the client fails to complete more than five items on a scale, the entire scale should be discarded, and the practitioner should learn why the scale was not completed before attempting to use it further with the client.

When used in clinical settings and applications, the CMP scales are short enough that they can be manually scored by the therapist. Each scale can be completed in three to five minutes, and each can be manually scored in about the same length of time. Figure 2–1 presents an example of a completed and manually scored GCS scale in which the respondent failed to complete three items. For that scale it can be seen that $\Sigma Y = 34$, and $N = 22$ so

$$
\begin{aligned}
S &= (\Sigma Y - N)(100)/[(N)(4)] \\
&= (34 - 22)(100)/[(22)(4)] \\
&= (12)(100)/88 \\
&= 1200/88 \\
&= 13.6
\end{aligned}
$$

It should be noted that the therapist circled each item that had to be reverse-scored and that each of the original scores for those items was clearly crossed out. Experience has shown that these two procedures help to reduce computational mistakes when manually scoring the CMP scales. However, it is very important to always recheck all item reversals and computations.

Figure 2–2 presents an example of a completed and manually scored CAM scale in which the respondent completed all 25 items.

FIGURE 2-1

GENERALIZED CONTENTMENT SCALE (GCS) Today's Date ________

NAME ______________________________

This questionnaire is designed to measure the degree of contentment that you feel about your life and surroundings. It is not a test, so there are no right or wrong answers. Answer each item as carefully and accurately as you can by placing a number beside each one as follows:

1 Rarely or none of the time
2 A little of the time
3 Some of the time
4 A good part of the time
5 Most or all of the time

34 − 22 = 12; 12 × 100 = 1200; 22 × 4 = 88

S = 1200/88 = 13.6

Please begin.

1. I feel powerless to do anything about my life. 1
2. I feel blue. 1
3. I am restless and can't keep still. 2
4. I have crying spells. 1
5. (circled) It is easy for me to relax. 2
6. I have a hard time getting started on things that I need to do. ___
7. I do not sleep well at night. 2
8. (circled) When things get tough, I feel there is always someone I can turn to. ___
9. (circled) I feel that the future looks bright for me. 2
10. I feel downhearted. 1
11. (circled) I feel that I am needed. 1
12. (circled) I feel that I am appreciated by others. 2
13. (circled) I enjoy being active and busy. 2
14. I feel that others would be better off without me. 1
15. (circled) I enjoy being with other people. 2
16. (circled) I feel it is easy for me to make decisions. ___
17. I feel downtrodden. 1
18. I am irritable. 2
19. I get upset easily. 2
20. It is hard for me to have a good time. 1
21. (circled) I have a full life. 1
22. (circled) I feel that people really care about me. 2
23. (circled) I have a great deal of fun. 2
24. (circled) I feel great in the morning. 2
25. I feel that my situation is hopeless. 1

N = 22 SUM = 34

5, 8, 9, 11, 12, 13, 15, 16, 21, 22, 23, 24

There it can be seen that $\Sigma Y = 83$ (after appropriate items have been reverse-scored), so the total score is computed as

$$\begin{aligned} S &= \Sigma Y - 25 \\ &= 83 - 25 \\ &= 58 \end{aligned}$$

FIGURE 2-2

CHILD'S ATTITUDE TOWARD MOTHER (CAM) Today's Date ________

NAME ______________________________

This questionnaire is designed to measure the degree of contentment you have in your relationship with your mother. It is not a test, so there are no right or wrong answers. Answer each item as carefully and accurately as you can by placing a number beside each one as follows:

1 Rarely or none of the time
2 A little of the time
3 Some of the time
4 A good part of the time
5 Most or all of the time

Please begin.

S = 83 − 25
= 58

1. My mother gets on my nerves. 3
(2.) I get along well with my mother. 4
(3.) I feel that I can really trust my mother. 2
4. I dislike my mother. 3
5. My mother's behavior embarrasses me. 3
6. My mother is too demanding. 4
7. I wish I had a different mother. 3
(8.) I really enjoy my mother. 3
9. My mother puts too many limits on me. 4
10. My mother interferes with my activities. 4
11. I resent my mother. 2
(12.) I think my mother is terrific. 3
13. I hate my mother. 2
(14.) My mother is very patient with me. 5
(15.) I really like my mother. 3
(16.) I like being with my mother. 4
17. I feel like I do not love my mother. 3
18. My mother is very irritating. 4
19. I feel very angry toward my mother. 3
20. I feel violent toward my mother. 3
(21.) I feel proud of my mother. 3
22. I wish my mother was more like others I know. 4
23. My mother does not understand me. 4
(24.) I can really depend on my mother. 4
25. I feel ashamed of my mother. 3

N = 25 SUM = 83
2, 3, 8, 12, 14, 15, 16, 21, 24

INTERPRETING THE CMP SCALE SCORES

Each one of the personal and social problems that is measured by the CMP scales is conceived as an intensity, severity, or magnitude continuum, and each scale was designed as a measure of the magnitude or severity of the problem that is being investigated.

Thus, if one person has a GCS score of 32 and another has a GCS score of 44, it will be concluded that the second person is more depressed than the first (assuming no errors of measurement). A similar interpretation is made with respect to each of the CMP scales.

The CMP scales were conceived and structured as measures of the degree, severity, or magnitude of personal and interpersonal relationship *problems* and not as measures of health or well-being. There are two reasons for this. First, it is difficult to define and quantify the concept of *health* directly. (It helps little to define health as the absence of such problems.) If health is defined as the absence of personal and interpersonal relationship problems, emphasis is shifted toward improving health and not toward the elimination, reduction, alleviation, or solution of problems. This leads to the second reason for structuring the CMP scales as measures of problems. That is, almost all forms of clinical therapy are primarily motivated by problem solving. Clients are accepted in therapy almost exclusively because some personal or interpersonal relationship disorder is identified as the focus or target of treatment, and the CMP scales were designed to measure and characterize the degree or severity of such problems. Thus, in using the CMP scales, a high score is always taken to indicate a more serious problem than would some lower score.

A critical question arises: How low must a client's score be in order to conclude that there is no clinically significant problem in the area being measured? Must the client's score be zero? Indeed not! An important feature of the CMP scales is that each of them has the same clinical cutting score of 30. That is, if a person scores above 30 on any of the CMP scales it is almost always found that the person has a clinically significant problem in the area being measured while persons who score below 30 are generally found to be free of such problems.

The clinical cutting score of 30 is useful for two reasons. First, it enables the practitioner to use the CMP scales as a rough diagnostic tool or indicator, but perhaps more important it represents a very useful treatment criterion. As a diagnostic tool, any one or several of the CMP scales might be used as screening devices to determine whether a person has a clinically significant problem in a specific area, and how serious that problem might be. However, it must be carefully noted that none of the CMP scales were designed to shed any light whatsoever on the source, origin, or cause of the client's problem, and they must not be used for such purposes.

Some users may be tempted to examine the client's responses to specific items on one or more of the scales to learn more about the nature or type of the client's problem. The scales were not designed for such uses; they will not produce consistent and reliable answers to such questions; and their use for such purposes must be avoided. If a client is having a problem, say, with depression, the client will usually make that known to the therapist. Thus, in using the CMP scales as diagnostic aids, their greatest utility arises from their ability to show the degree or magnitude of the problem and not whether a problem is present. Moreover, if the therapist wants and needs information about the nature, source, origin, or type of problem, there are much better sources for this kind of information that are available: the client, relatives, coworkers, friends, and the therapist's analysis of the problem. In short, therapists can and should use the CMP scales as diagnostic aids because they do a reasonably good job of showing whether a client has a clinically significant problem and how serious that problem may be, but none of the scales should be used in any other diagnostic context.

As a treatment criterion, the clinical cutting score of 30 provides the therapist with a very convenient goal or standard for evaluating treatment. For example, if a client began treatment with an IPA score of 54 and eight weeks later the client scores 13 on the IPA, there is little justification for continuing to treat that specific parent-child relationship problem because the score is far below the clinical cutting score of 30. However, if a client begins treatment, say, with an IMS score of 61 and 14 weeks later obtains an IMS score of 44, the therapist can claim only that the problem has been reduced but not solved. Although the therapeutic goal can be seen as one of resolving the client's problem to the point where the client will score below 30 on the relevant scale or scales, how far below 30 the score must go before treatment is terminated is difficult to specify. A decision in this regard is nearly always one that is negotiated by the therapist and the client.

Although the score of 30 has been found to be a useful and important cutting point for understanding and interpreting the scores obtained from the scales (it can be used as a criterion, benchmark, or goal), it is not a perfect indicator, and one should not rigidly adhere to it. No scale is perfect, and no cutting score that is associated with clinical scales is perfect. Some clients will obtain scores in excess of 30 when it is impossible to show by any other evidence that they have a clinically significant problem in the area being assessed; such outcomes are rare, however. Likewise, some clients will score below 30 when there is other unmistakable evidence to show that they have a clinically significant problem; these are also rare outcomes. There is currently no measurement scale of the type represented by the CMP scales that is so good that one can use it in clinical practice and rely completely on its technical performance characteristics.

In using the CMP scales in clinical practice, it is absolutely essential that obtained scores be carefully evaluated in relation to all other data and information that is available about clients and their problems.

If a client scores above 30 on one of the CMP scales and all other clinical evidence clearly indicates the client does not have a problem, the proper conclusion must be that the client does *not* have a problem. If a client scores below 30 on one of the CMP scales and all other clinical evidence indicates the client has a problem, then the proper conclusion is that the client has a problem. This kind of clear and unambiguous incongruence between obtained scores and other types of clinical evidence is quite rare, but it happens occasionally, and the therapist must be alert to it. There are two basic reasons for such incongruity. The first concerns the possibility of a social desirability response bias or the possibility that the client is lying, and the second concerns the measurement error inherent in the scale. I will now discuss these reasons.

The standard error of measurement or SEM for each of the CMP scales is roughly five or six points. This means that if a client's score changes from one week to the next by plus or minus five or six points or less, one should regard such a change as inherent noise or error in the measurement scale and not as evidence of real change in the degree or severity of the problem. However, if the client's score changes by more than five or six points in either direction one can have considerable confidence in the conclusion that real change has occurred. If the score changes by an amount greater than twice the SEM (see Chapter 6 for precise SEM values) on any of the CMP scales, one can be about 95 percent confident that real change has occurred in the degree or severity of the client's problem.

This measurement characteristic of the CMP scales has a great deal to say about the use and interpretation of the cutting score. First, it is now apparent that the SEM makes the cutting score of 30 an imperfect indicator of the presence or absence of a problem. Second, if a boundary of plus or minus five points (or the SEM value for a particular scale) is placed around the score of 30, one thereby defines a *zone* of ambiguity. That is, if a client obtains a score on one of the CMP scales that falls between 25 and 35, the therapist may want to obtain additional supporting evidence before concluding that the client does or does not have a clinically significant problem in the area being assessed.

Thus far no clinical case has been seen in which the client clearly acknowledged a problem in the domain being measured and obtained a score lower than 20 on any of the CMP scales. Neither has any case been seen in which a client unmistakenly was free of a problem in the domain being measured and who scored over 40 on any of the CMP scales. These scores represent a boundary of about plus or minus twice the SEM around the clinical cutting score. In most cases a score of 30 can be used as an effective cutting point; in ambiguous situations where complete diagnostic information is not available, it is wiser to use a cutting *range* of 25 to 35 for each of the scales. More precise ranges can be defined by using the SEM values reported in Chapter 6.

When a client provides socially desirable responses, engages in other forms of favorable impression management, or deliberately falsifies responses to the items on the CMP scales to minimize the problem, an important and significant error is introduced in the measurement of the client's problem. A similar type of error is encountered when a client provides item responses that make the problem appear to be more serious than it really is; this is referred to as a form of malingering regardless of whether the behavior is deliberate or not. Social desirability and malingering responses contribute to the size of the SEM discussed previously, and in the final analysis there are only a few things that the therapist can do to minimize or control these two types of measurement error.

Of utmost importance, the therapist must clearly explain and demonstrate to the client that the scales will be used for the client's benefit and that the therapist has a valid and legitimate right to have the information that is being sought. The therapist must also emphasize to the client the absolute necessity for providing the most accurate responses possible. Although the therapist has no other means of controlling these types of errors, there are other ways of detecting and accounting for them.

The most effective method for validating the client's scores on any of the CMP scales is to assess them in relation to numerous other items of information about the client's problems. If a 14-year-old boy obtains a score of 47 on the IPR, he should be able to indicate the specific types of problems he is having with his peers, and his parents, friends, and teachers should be able to provide corroborating evidence that a problem in this area really is present. If such evidence cannot be obtained, the therapist should suspect the presence of malingering, check it out, and try to determine the reasons for it if it is confirmed. Although a woman may be severely depressed and seek help, she may initially provide a socially desirable score on the GCS of, say, 31, because she does not want to face or confront the therapist or herself with the true magnitude of her problem.

The purpose of this chapter has been to provide specific instructions for scoring and interpreting client responses to the CMP scales. However, additional discussion concerning both topics can be found in later chapters of this manual.

3 Administering and using the CMP scales

Successful use of the CMP scales in clinical settings depends primarily on the skill of the therapist in administering and using the scales for specific purposes; this chapter provides some practical guidelines in these areas. Although skill and sensitivity are required for successful use of the scales, these are quickly and easily acquired on the job and do not require specialized technical training. Further details concerning the interpretation of scores obtained from the scales are also provided.

ADMINISTERING THE SCALES

One of the worst possible things the therapist, worker, or counselor can do in administering any of the CMP scales is to convey to the client the impression that the therapist is uncertain or unsure of the value or importance of administering the scales or obtaining the client's responses to them. For whatever reason, some therapists who have never used measurement tools of any sort in clinical practice seem to feel they must apologize when asking the client to "take a test," and an apologetic attitude on the part of the therapist immediately conveys to the client an impression that completion of any questionnaire or scale is an unnecessary, unwarranted intrusion upon the client and the therapist. This must be avoided.

The CMP scales are *not* tests. There are no right or wrong answers to any of the items on any of the scales. The scales are designed to obtain information about the severity or magnitude of the client's problem in one or more areas, and the therapist's need for such information constitutes a legitimate basis for asking the client to complete one or several of these measurement tools.

Unfortunately, many clients are wary and distrustful of psychological "tests," and some may be unwilling to fill out any scale that is not properly explained to them. Equally unfortunate is the fact that there is a real basis for client fears and concerns about completing psychological tests. Patients, clients, research subjects, and ordinary citizens are occasionally asked to take psychological tests whereupon the scores and their meaning are hidden away, protected, left unexplained, used to attach unflattering diagnostic or descriptive labels to the respondent, or are otherwise beclouded in an aura of secrecy and mystery. It is therefore not difficult to understand why the public has developed a sense of paranoia and suspicion about psychological tests or scales.

To properly use the CMP scales, the therapist must overcome these fears and concerns of the client (as well as the therapist's own concerns in this regard), and it is nearly always surprisingly easy to do by following a reasonable set of guidelines in administering the scales. The remaining sections of this chapter describe procedures or rules that have been worked out in clinical uses of the scales and thus far have proven to be highly effective. Basically there are only three such rules or procedures: understand how the scales perform; assure the client of their importance; and explain their purpose and how they will be used.

FAMILIARITY WITH SCALE PERFORMANCE

It is extremely important that the therapist have a basic understanding of how the CMP scales work and perform before administering them to a client. How does one gain such experience if one has never administered them? Very simple; take them yourself. This is a very important exercise that represents one of the principal means whereby therapists can train themselves to better understand and interpret scores that will be later obtained from clients. Therapists who have never used the CMP scales should go through the entire package and complete every scale that is applicable to their own personal life situation. When therapists complete the CMP scales, they should carefully examine their scores in relation to the way they feel about their own affective responses to each area that is measured. That experience will provide an important understanding of the meaning of scores that will later be obtained from others. For example, if one feels a bit gloomy (mildly depressed), completes the GCS, and obtains a score of, say, 23, it will then be possible to develop some appreciation or understanding of the level of the depression someone might feel who obtains a score of 44 or 60.

Few of us have been so fortunate as to have escaped ever having any of the problems represented by the CMP scales, and these personal problems can be used by therapists both to test out the scales and to develop some appreciation for and understanding of obtained scores. Suppose, for example, that at one point in your marriage you were having serious relationship problems with your spouse. You might try to recall those old feelings and then complete the IMS scale in terms of the way you felt about your marital relationship at that time. By comparing the way you felt about your marital relationship with the obtained IMS score, you can gain considerable insight into the way a client may be feeling who obtains a similar score. No doubt, it is unpleasant to dredge up old wounds, hurts, and problems from the past, but the dividends will be important in terms of understanding the CMP scale scores that are gotten from clients. As an aid to helping the therapist gain experience and familiarity with the CMP scales, a number of self-training exercises have been provided in Appendix A. All of those exercises that are applicable should be completed by those who are not familiar with the use and scoring of the scales.

ELICITING THE CLIENT'S CONFIDENCE

If therapists demonstrate or suggest by their attitude or manner that they are uncertain of the value or importance of the CMP scale responses, the client is also likely to have little confidence in them, and the client may then refuse or be reluctant to complete any of the scales. If that happens, the therapist may have lost an important opportunity to gain information that could help to understand the severity of the client's problem. Also, the door may then be closed to an opportunity to determine more precisely whether one's treatment efforts were or were not effective in helping the client to solve the problem.

The best way to avoid these problems is for the therapist to demonstrate to clients a high degree of confidence in the value and importance of what they do together. Therapists should never be apologetic or show hesitation when asking a client to complete one or more of the scales. After all, clients seek help because they feel they cannot handle their problem alone, and they neither want nor need to believe that therapists are unsure of themselves, their tools, or their methods of gathering information or providing treatment.

EXPLAIN THE PURPOSE OF THE SCALES

Clients have a right to know why you are asking them to complete one or more of the CMP scales, and they have a right to know how therapists will use the results. So tell them. By doing so, the therapist shows respect for the client and is likely to get better results. Moreover, by clearly and adequately informing clients of the purpose and use of the scales, they are included in the planning and conduct of the treatment. They are less likely to feel they are strangers in some mysterious world through which only the therapist, as expert, has right of passage.

When providing this information it is not necessary to give lengthy, detailed explanations, and clients certainly do not need details concerning the psychometric properties of the scales. For example, near the end of the first intake interview the therapist might say to a client,

> Mr. Johnson, you have made it clear to me that you and your wife are having some marital problems and that you are feeling depressed about that. However, I want to nail this down a little better and I want you to fill out two short scales for me (the IMS and GCS). When you finish the scales I will score them, show you the results, and tell you what they mean.

If the client asks what the scales are supposed to do or reveal, by all means tell him. You might say something like,

> The first scale, the IMS, will show us how severe or serious is the marital problem, as you see it, and the GCS scale will tell us a great deal about how depressed you are feeling.

The keynote to successful use of the scales is to be as open and candid about their use and the results as good clinical judgement will allow. For example, a teenage boy might score, say, 93 on the CAF scale. The therapist certainly does not need to speculate aloud about the possibility that the client may physically attack his father. Instead, the therapist might say, "John, you scored 93 on the scale, and that clearly matches up with all the difficulties you've told me that you're having with your father."

HANDLING CRISIS AND PANIC

At the beginning, and during the conduct of treatment, therapists often encounter severe crisis or extremely intense emotional situations—including a variety of panic reactions. When these arise, the therapist's major concern is to deal with the crisis, and at such times it is unwise to ask a client to fill out the CMP scales. To do so would likely convey to the client an impression that the therapist either does not understand what is going on or does not care—or both. When a severe crisis arises it is best to suspend use of these or any other scales and return to them only after the crisis is over or well in hand.

WHEN TO ADMINISTER THE SCALES

The question often arises as to when is the best time to administer the scales and how often. There are no well-established rules in this regard, and the frequency and timing of the use of any of the scales must be judged in relation to therapists' need for information. Generally it has been found that it is best not to ask clients to complete any of the scales at the beginning of an intake interview or before they have seen a therapist. For intake interviews, it is probably better to administer the scales near the end because the therapist will have a better idea of which ones are most useful for that particular case and which ones are definitely not appropriate. However, there are no fixed rules in this regard. Some clinics may choose to use several of the scales as screening devices by asking clients to complete the scales before seeing a therapist. Properly handled by a skilled nurse or receptionist, the scales can be used in this manner, but in most cases the best results are probably obtained by having the therapist introduce and explain the use and purpose of the scales within the context of the client's problem and treatment.

Once the intake process is finished, the therapist may wish to administer the scales on a regular or periodic basis to monitor and evaluate the client's progress in treatment; this is probably the most powerful use of the scales (an entire chapter has been devoted to this function). Experience has shown that the scales perform very well even when they are used daily. However, such frequent use is not recommended for two reasons: It is extremely rare that such close monitoring is ever called for, and if it is not, the client will become bored and irritated with such usage.

In most clinical situations, it is not a good procedure to administer the scales more than once a week, and a generally satisfactory result is usually obtained by administering the scales once every two weeks; biweekly administration of the scales should probably be regarded as a standard operating procedure. One should never adhere blindly to such a standard, and therapists must develop an administration schedule that is designed to meet their needs for information about client change.

For example, if one establishes an implicit or explicit short-term treatment contract in which treatment will be terminated within eight weeks, the scales should probably be administered every week during treatment. For a middle range treatment regimen that is expected to run for 12 to 20 weeks, the scales should probably be administered biweekly. Longer treatment periods might use a three- or four-week scale administration schedule. The most general and effective rule or guideline is to administer the scales with sufficient frequency to maximize receipt of useful information about the client's progress without inducing boredom on the part of the client.

WHICH SCALES TO ADMINISTER

In the overwhelming majority of the cases seen in clinical settings, it is very easy to determine which scales to use with any specific client. The best rule is to use the scale or scales that relate directly to the major problem that is to be treated. Thus, if a client is being treated for depression, the GCS should be used; if the client is seeking help with a marital problem, use the IMS, and so on. In short, use good clinical judgement in selecting appropriate scales.

Some of the scales go together quite nicely as pairs or even triples. For example, in the general population there is a fairly high correlation between depression and problems with self-esteem. Thus, a client who begins treatment for depression should probably complete both the GCS and the ISE scales. Clients who are moderately or severely depressed often have self-esteem problems as well, but one encounters some cases wherein the client is very depressed but has no apparent or measurable self-esteem problem. The reverse is also true. If a client is being treated for a severe self-esteem problem, both the GCS and ISE scales should be completed.

If a client or couple complains of a problem in their marital relationship, the IMS scale should be filled out. However, marital problems are very frequently accompanied by sexual problems, and in such cases it is wise to consider use of the ISS scale as well. Again, the reverse is also true. Treatment of a sexual relationship problem should involve both the ISS and IMS scales.

When treating a child for a relationship problem with a parent, the therapist should check out the possibility of a similar problem with the other parent. Thus, the CAM and CAF scales are usually given together. Also, if the child indicates a relationship problem with the mother, the therapist should try to find out how the mother feels about her relationship with the child by having her complete the IPA scale. As a check, the father should also complete the IPA. Moreover, in many such cases, it would be advisable to have mother, father, and child complete the IFR scale. The combined information or scores from all these sources will aid the therapist in identifying which components of the family are most problematic and may help to develop a treatment focus.

A few cases have been seen in which there was an obvious need to use a particular scale that was totally overlooked by the therapist. In one case, an adolescent was being treated for depression, and the therapist was using the GCS scale. One look at the GCS scores clearly indicated the youth was very depressed, and when asked why the client was depressed, the therapist replied that the client was not able to get along with his mother. When the therapist was asked to show the client's CAM scores, the therapist said that the CAM scale was not being used: The client's mother had not been asked to complete the IPA scale. In other words, the therapist was treating a relationship problem but measuring the level of depression. It would have been much better to monitor both by using the GCS, CAM, and IPA scales. Oversights of this sort can sometimes be very costly in terms of failing to recognize real change and in terms of obtaining misleading or otherwise inappropriate information.

AVOID EXCESSIVE ADMINISTRATION

It would obviously be impossible to ask any client to complete all nine of the CMP scales every week. The client would not do it, and if it was done there would be no time left in the hour to conduct treatment. However, if the therapist is trying to monitor three separate problems, it will take the client about 15 minutes to complete the three CMP scales. That is a lot of time to use in a therapeutic session. This can be handled in many cases by asking the client to complete the scales at home or by having the client arrive a few minutes early for a particular session. The client can obtain the scales from a receptionist, and the receptionist or the client can be easily taught to score the scales.

The major point, however, is that the therapist should not use any scale that does not deal directly with the problem or problems being actively treated or worked on. To do otherwise represents an unnecessary imposition on the client and an intrusion on valuable clinical time that could be used more productively. Another way to reduce the amount of time used to administer and score the scales is to stop using a particular scale once the client's problem in that area has been resolved.

CHECK OUT INCONSISTENT RESPONSES

The therapist must always be on the alert for the possibility of receiving incorrect information regardless of whether that comes from verbal responses by the client or from the client's responses to the scales. One of the best checks on this comes from a comparison of the two. If a client complains of a severe problem in a relationship with a peer group but scores below 30 on the IPR, something is likely amiss, and the therapist must not ignore it. If the client says there is no problem with depression but scores well above 30 on the GCS, the conflict of information constitutes a problem that must be examined. The client in either of these examples may not be giving accurate verbal responses or may be giving inaccurate responses to the scales. The therapist must determine which is the case. In rare instances it may be found that one or more of the CMP scales is simply not valid—it just will not work—with a specific client. If that occurs, the wisest choice is to immediately terminate use of that particular scale. However, it has been found that the CMP scales are sufficiently dependable, that one should not leap to the conclusion that the problem resides with the scale. In fact, the scales should be abandoned *only* after other explanations have been thoroughly checked out and one has convincing evidence that the scale is not going to do the job it was designed to do with a specific client.

SOCIAL DESIRABILITY AND MALINGERING

A single reading of any of the CMP scales will make it apparent that clients can easily make themselves look as good or as bad as they wish on any of the scales, and to partially control such responses, this should be discussed at the beginning. It is not necessary to dwell on this problem or to give a lengthy discussion. The procedure recommended here is to caution the client the first time one of the CMP scales is completed. One might say, for example,

> Mr. Edwards, when you fill out this scale you will see that you can make yourself look as good or as bad as you wish, but if you do that you will not help yourself or me. What I want you to do is give me the most accurate answers you can to every item on the scale. If you do that, you will see that the scale will be very helpful to us both.

By and large, such simple instructions have been sufficient to ward off social desirability and malingering responses in clinical settings.

LYING AND LIE SCALES

In spite of one's best efforts to develop a trusting relationship with a client and to obtain accurate information about the client's problem, some clients will provide incomplete or misleading information. Some will simply lie to you. The reasons for such behavior are numerous and complex and will not be discussed in this manual.

The topic is relevant in terms of using the client's responses to the CMP scales.

Although one definitely wants to avoid and minimize social desirability, malingering, and lie responses on the CMP scales (and in discussions with the client), it is generally inadvisable to dwell on this problem when working with voluntary clients who seek help. It is terribly important to face squarely the fact that clients are individuals with an undeniable right to privacy, and they may choose to protect that privacy by lying to you until they feel they can trust you with material they feel is sensitive. In short, if clients feel they do not want to reveal their problem, or information about that problem, they have a clear right to withhold it. Their therapeutic contract with a therapist does not in any way abridge their civil rights.

For these reasons, it is strongly recommended that so called "lie" scales should not be used in clinical practice as a means of checking the accuracy of the CMP scales. If clients do not want a therapist to know certain information about themselves or their personal situation, it is extremely difficult to trick them into revealing it by using a so-called lie scale; efforts to do so may generate an attitude of distrust that worsens rather than reduces the problem. If a client is lying and that is preventing any productive therapeutic work, it is better to confront the client and deal with the problem openly and honestly. Trickery on the part of the therapist is all too likely to generate trickery by the client.

This problem was rather forcefully illustrated by a young man who sought help with a problem he was having in his relationship with the woman he was living with. He gladly completed the IMS scale, but when asked to complete the ISS scale, he told the therapist that was not necessary because he and his partner had a great sex life together. The therapist asked him to complete the ISS as a check. The client scored in the mid-40s on the IMS and scored 13 on the ISS. This pattern was repeated over the next two weeks as shown in Figure 3–1. During the fourth interview the client scored in the mid-40s on the IMS scale, as before, but the client's ISS score rose to 52. When the therapist saw this, he pointed out to the client that the ISS scores had been below 20 for the first three weeks and now jumped to more than 50. The therapist asked for an explanation and the client replied, "Well, I've worked with you for about a month now and I feel like I can now trust you with this information." The problem had been there all along, but the client could not reveal or deal with it until he felt secure in his relationship with the therapist. Clients check out their therapists very thoroughly; it takes time to do that, and they have every right to do so.

FIGURE 3-1

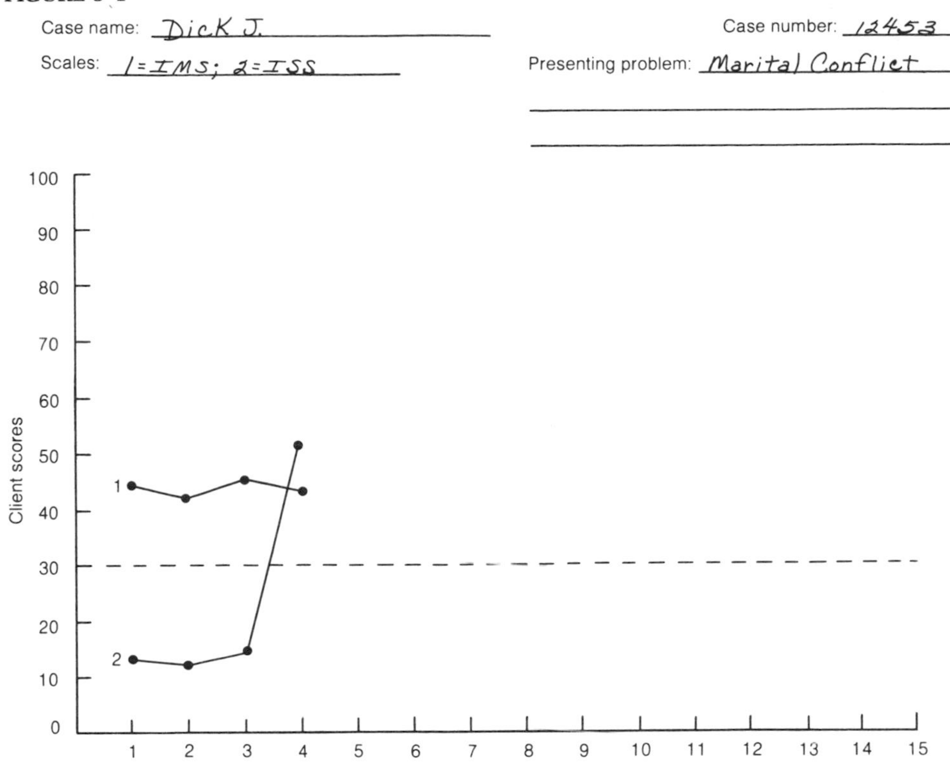

USES FOR THE CMP SCALES

There are many different ways in which the therapist or research investigator might use the CMP scales. Some of those are discussed in this section, and once the therapist has gained experience in using the scales other kinds of applications will come to mind. The practical limits for their use are a function of both the situation and the user's imagination and creativity.

Diagnostic aids

The use of the CMP scales as diagnostic aids has already been commented on, and that discussion will not be expanded significantly. As cautioned previously, the scales do a poor job of shedding light on the source or origin of the client's problem, and their primary use is to measure the degree, severity, or magnitude of the client's problem. As diagnostic aids, the scales can be used singly or in combinations. Under well-controlled conditions they can be used as diagnostic screening devices but such usage should be undertaken with care.

From a diagnostic point of view, several of the scales can be used to reflect the degree of congruence or incongruence between individuals with respect to a specific problem area. For example, a husband and wife who score, say, 42 and 24, respectively, on the IMS may be quite different from another couple who score 48 and 51. The score distances between partners on many of the scales can provide the therapist important clues about how to analyze the case and how to structure treatment. Such score differences often surprise the client and the therapist.

Evaluation tools

The most powerful use of the CMP scales arises in the context of monitoring and evaluating the course of treatment with an individual, couple, family, or small group. By administering one or more of the scales at regular or periodic intervals, the therapist can readily determine whether the client's problem is decreasing, remaining fairly stable, or getting worse. Such information can be extremely useful in the conduct and modification of a treatment regimen. This usage of the CMP scales is so important that an entire chapter has been devoted to it, and no further comment will be made here.

A communication device

In a number of cases, the CMP scales have served as important tools of communication among spouses and family members. One couple, married about 20 years, completed the IMS scale and felt they had learned nothing new about their marriage when they saw their scores. However, they exchanged scales and examined each other's responses to the items on the IMS. By examining the way their spouse responded to each item they each expressed some surprise and in several instances said, "I didn't know you felt this way about. . . ." In short, their usual patterns of communication were not enabling them to accurately perceive how their partner felt about specific aspects of their marital relationships.

It is not recommended that the items on any of the CMP scales be interpreted by the therapist or researcher as a set of subscales. However, when working with interpersonal relationship disorders, it is often very useful to have each partner examine the other's responses to the individual items on one or more of the CMP scales. For example, a husband may be having some difficulty understanding the wife's rather serious self-esteem problem, and if a treatment confidence is not violated, it may be very helpful for him to examine his wife's responses to the items on the ISE scale. Also, to reinforce positive gains in treatment, he may be asked to do this at the beginning of treatment and then much later—after improvement has become apparent. By comparing item responses at two different points in time, both partners may then have a much clearer idea of specific areas of improvement.

Projective measurements

Although none of the CMP scales were designed for use as projective tests, they can be used as such under certain specialized circumstances. Suppose, for example, that a therapist is working with a man who has a serious problem with his 16-year-old daughter. The therapist will likely have the father complete the IPA scale in order to measure the degree or magnitude of the father-daughter relationship problem. The therapist also may want to know how serious the father thinks the problem might be from his daughter's point of view. This can be determined by simply asking the father to complete the CAF scale the way he *thinks* his daughter would do it. Such uses of the CMP scales can be very important. If the father's CAF score is much greater than his IPA score, the therapist might explore the possibility that the father sees the problem as being largely attributable to his daughter. If his IPA score is much larger than his CAF score, he might be seeing himself as the locus of the father-daughter relationship problem.

In other words, there may be many instances in which it would be useful to have clients complete one or more of the CMP scales the way they *think* some other person might respond. In these cases, the respondent is projecting his or her perception of another person's view of the problem. Again, none of the CMP scales were designed for such usage and one must be cautious in interpreting data obtained from such applications. Under special circumstances, however, such data can be an important aid to both the therapist and the client.

Measuring misperceptions

In some cases a relationship problem can develop, be maintained, or worsen in part as a function of each person's misperception of the relationship. If the therapist can help to identify the magnitude of such misperceptions, that alone may prove to be a powerful therapeutic device. Several years ago a mother complained that her husband was acting abusively toward their adolescent daughter, and the mother's complaints were a source of friction between the parents. The CMP scales were used to obtain some measure of the misperception of these relationships. The father completed the IPA to show how he felt about the daughter; the daughter completed the CAF to show how she felt about her father; the mother completed the IPA the way she *thought* the father would respond to the daughter; the mother completed the CAF the way she *thought* the daughter would respond to the father; and the father completed the CAF the way he *thought* the daughter would respond to him. By comparing the scores for the three family members, the therapist was able to show that the mother severely misperceived the father-daughter relationship problem, and, as a consequence, she markedly reduced her complaints about the father's behavior. An unexpected result in this case was that the father had also misperceived his relationship with his daughter; the daughter felt much more positive toward her father than he thought she did. No therapeutic intervention was used other than the identification of these relationship misperceptions, and a follow-up with this family 15 months later showed that the problem had abated.

4 Some tough questions

Over the past few years many people have asked a number of specific and important questions concerning the CMP scales. Because these questions arise repeatedly, it is useful to state them explicitly and respond to them. These materials are presented in this chapter in the hope that this question-and-answer format will help users to make better decisions concerning the administration and interpretation of the CMP scales.

How do I know whether a client has provided truthful and accurate responses to one or more of the CMP scales?

If no further information is available, it is impossible to be sure. Trusting the responses then becomes an act of faith. However, in clinical practice the therapist nearly always has a great deal of information about the client, and that must be used to assess all of the scale responses. It must be remembered that this question must be asked for virtually all of the information that clients provide therapists—whether it comes from responses to the scales, questions raised by the therapist, or is volunteered by the client.

The major tools that therapists have for dealing with this issue are consistency and verification. Another consists of therapists' judgements about the type and quality of their relationships with their clients. If clients present a consistent and convincing picture of someone who is willing to disclose the more painful aspects of problems they seek help with, therapists can then be much more certain of the quality of responses given to the CMP scales. If clients tend to deny, hide, or diminish the painful realities of the problem, it is reasonably certain that such behavior will carry over to some extent to the clients' responses to the CMP scales. Ironically, however, there have been numerous instances in which clients have been much more candid in their responses to the scales than in verbal accounts of their problems.

Generally speaking, the scales perform very well because many clients take the work of therapy rather seriously and provide candid responses. Moreover, even when no other information is available about the client's problem, reliance on the CMP scales is not an act of blind faith. After all, the scales are highly reliable and valid, and they generally provide accurate and dependable information.

Quite often I have noticed that scores on these scales are large enough to suggest that the problem is much more serious than I would have suspected on the basis of the client's discussion of the problem. What does this mean, and why does it happen?

This phenomenon usually occurs early in the treatment relationship and there are good reasons for it. First, many clients are embarrassed or reluctant to verbally disclose the full force of their feelings to a person with whom they do not yet have a close relationship. Second, therapists are not usually adept at judging the severity or magnitude of client problems. This is not to criticize therapists but simply acknowledges the fact that therapists cannot get into someone else's head and accurately perceive how that person may be feeling. This is one of the major benefits of the scales. They provide therapists an improved ability to assess the severity or magnitude of the client's problem.

What should I do if I strongly suspect that a client is not giving truthful or accurate responses to the scales?

The first thing to do is to give little, if any, credence to the scale scores. If the problem persists, it may be necessary to terminate use of the scales altogether: Why collect data about which there can be no confidence? The answer to this question deals more with treatment than it does with measurement. The therapist may have to confront the client in one way or another with a suspicion about dishonesty. This can be risky and therapists should be confident of their assessments. Perhaps the best strategy at first is to confront the client gently with the fact that the information given by the client is either not consistent or that it does not match information obtained from other sources. Sometimes the client will provide a useful and reasonable explanation; sometimes the problem will then smooth out; and sometimes the therapist may have to assess whether continued work with such a client would be warranted. In rare cases the focus of treatment may shift away from some initial presenting problem and become focused on a problem that is represented by a pattern of lying and distortion. In most cases, clients have a legitimate reason (at least in their own minds) for not providing certain kinds of information, and it is the therapist's job to discover this and to take appropriate, constructive action. The best policy is first to give the client the benefit of the doubt; check out the inconsistencies carefully; and then proceed on the basis of the best information that can be obtained.

If a client completes one or more of the scales on a regular basis, what does it mean if there is a sudden dramatic rise or decline in the total score?

It almost always means that some specific or special event has occurred that has had an impact on the client's problem. Such changes are vitally important clues that should be investigated thoroughly and noted in the client's record. The ability of the CMP scales to reflect such changes represents another major benefit that arises from their use.

The special events that are associated with sudden dramatic changes in client scores are sometimes very obvious or apparent to the client. the therapist, or both. Even so, do not make presumptions. Check out carefully any such changes, and verify the events that are associated with them. Sometimes the events that are associated with dramatic score changes are not obvious. Sometimes the client may be reluctant to discuss what may be happening, but it is important to carefully investigate any and all such changes. The events that are associated with dramatic score changes may occur outside of the treatment relationship, or the client may be responding to something that is taking place in the treatment situation. Some of these events, whether internal or external to the treatment situation, may be such that therapists can exercise some influence or control over them. If they are amenable to influence or control they may serve as an important mechanism through which the therapist can improve the likelihood of a more positive outcome for the client. If it is not possible to influence or control such events, knowledge of how they affected the client may provide very useful information about directing the remaining course of treatment.

What constitutes a "dramatic" score change for these scales?

Any sudden change of 10 or more points in the total score for any of the scales should be examined carefully. Keep in mind that the SEM for all the scales is about five points. That means that in the absence of any real change in the severity of the client's problem, the scores on any of the CMP scales should moderately fluctuate around a single stable level. In fact, the clear majority of all such scores will vary by about plus or minus three or four points. (For further details see Chapter 5.) It is very important to point out that dramatic changes in scores can be used in a planned and deliberate manner as a means of assessing and evaluating the effectiveness of therapists' treatment efforts. This use of the scales is discussed in detail in Chapter 5.

Can I administer any of the scales to very young children?

No. It is very doubtful that any of the CMP scales are reliable or valid for persons under the age of 12. Occasionally a therapist may wish to try using one or more of the scales with a very precocious 11-year-old, but even that should be done with caution. If it appears to work, then proceed accordingly. As a general rule, however, the successful use of the CMP scales requires the emotional, affective, and cognitive maturity of respondents that is rarely observed among those below the age of 12.

Can I use the scales with people who are illiterate?

Generally speaking, that can be done. If the contents of the scales are read to clients and they are helped to understand how to respond to each item, they can be used successfully with illiterates. Frequently, however, people who are illiterate will not understand some of the words in the scale items when read aloud. If the therapist can define or interpret such words so that the client will understand them, there should be no difficulties. If a client cannot understand a few items (up to five) those items may be omitted. However, it will then be necessary to score the scale as described in Chapter 2 for such cases.

If a literacy problem is even remotely suspected, it should be checked out carefully. In a rural mental health clinic a therapist was attempting to use one of the CMP scales with a man who was, by most middle-class standards, quite wealthy. The client's affluence led the therapist to assume that literacy was obviously no problem. The therapist complained to the agency administrator about the nonsensical pattern of the item responses, and the administrator explained that he knew the client personally and that the client did not even know how to write his name.

Can I interpret individual item responses on the scales?

This is risky, and it is generally not advised. However, there are clear exceptions, as the foregoing example illustrates. While the total scores for each of the scales have good reliabilities (.90 or better), individual scale items are always much less reliable. Many items have reliabilities of .40 or less (see Chapter 6 for details) and may, therefore, be misleading. In spite of these cautions, it must be said that inspection of item responses sometimes gives important clues about the nature of the client's problem and what to focus on in the treatment situation. Generally speaking, therapists should not depend heavily on separate item responses and they should be used or interpreted only after much experience has been obtained in working with all the scales.

If a client obtains a score that is slightly larger than 30, can I be certain the client has a clinically significant problem? If the score is slightly below 30, can I be sure there is no problem?

Such borderline scores can never be interpreted with great confidence. In such cases, it is always necessary to obtain additional clarifying information from the client and other sources. This is true especially when the client's score falls within a range from about 25 to 35, and when the client's score departs or moves away from that range the therapist can have much more confidence about a decision concerning the presence or absence of a clinically significant problem.

Bear in mind, however, that the ambiguity with respect to such borderline scores has to do with making a *decision* about the presence of a clinically significant problem. Remember that the scales were designed to measure the severity or magnitude of personal and social problems. Thus, for example, if someone obtains a score of 28, the therapist can be confident that the person is not completely free of the problem. We all suffer—more or less. However, most of us function and carry on without the need for professional intervention. Thus, if someone obtains a score below 30, the central question is not whether the person is or is not depressed but whether the depression is clinically significant.

There is high consensus within the professional community that therapists should not be working with clients who have no clinically significant problems. To do otherwise raises serious ethical questions, but that is not the most important issue here. The most important issue is to avoid the risk of overlooking or failing to attend to a clinically significant problem when consulted for help by a client. It is in this regard that the scales can be very helpful.

Can I use all or most of the CMP scales as a screening or diagnostic battery and have clients complete them during the intake or admission process?

Generally, that is not a good practice, and it is not encouraged. It would take a client about 45 minutes to complete all nine of the CMP scales, and that would surely be seen by most clients as an unnecessary, even an alienating, task. When people feel they are in trouble and they seek professional help, they almost always want to talk with a professional who is interested in hearing what they have to say about the reasons they seek help. Ancillary personnel should not ask clients to complete screening devices before the client has been seen by a therapist, and they should not do it after the intake interview unless that has been arranged for by the therapist and with the client's consent and cooperation. Further information about a general screening device can be found in Appendix B.

If I see that a client's scores are declining over a period of time, can I be sure that such a decline represents real growth?

Yes, in almost all cases. However, the decline must be large enough to exceed the SEM for the scale, and the amount of the decline nearly always indicates the amount of growth that has occurred. The ability of the scales to detect and reflect such growth over repeated administrations represents one of their most important benefits. They were designed specifically for such applications.

Bear in mind, however, that there are many reasons for declining scores. In some cases, clients may wish to terminate therapy, or they may, for several reasons, want to appear to be making progress when little or none has occurred. Therapists must always be alert to such possibilities. The best way to guard against being misled by false declines is to check the scale responses for consistency or congruence with all other clinical evidence that is available.

Oddly enough, the scales will sometimes detect real growth when the therapist cannot see it. In one case the therapist had a client complete the GCS scale on a regular basis but failed to compute total scores or to plot them graphically (see Chapter 5); the therapist focused only on the item responses. After a period of four weeks the therapist could see no progress and, in consultation with his supervisor, decided to transfer the case to another therapist. The client was puzzled by this decision, believed that he had made progress, and decided to terminate rather than transfer to another therapist. When the author was consulted about this case, he scored the four GCS scales that were in the client's record and found unmistakable evidence of improvement. The decision to transfer the case was undoubtedly misguided and probably was directly responsible for a premature termination of treatment. The proper use of the scales may have prevented that.

The case example you just cited makes it appear that when data from the scales contradict the therapist, you should abide by the scales. Is that what you're saying?

Not at all. The point is that the therapist often must make decisions in the absence of reliable data. The scales can partially reduce that problem. In the case just cited, the therapist failed to use reliable data that he had obtained by administering the scales. Over the years it has been routinely observed that therapists do a very poor job of predicting a client's score on any of the scales—even for clients they know very well. However, once they see a client's score on any of the scales, they rarely dispute its validity. At least one reason for this phenomenon is that therapists have the ability to confirm a score's validity but not to predict its value in advance, and most of them know this. Again, this is not a criticism of therapists but an acknowledgement that it is *very* difficult for anyone to predict any of the CMP scores no matter how well they may know the client.

Keep in mind, because the scales are self-report devices, they are better able to "get inside the client's head" than is the therapist. More correctly, the scales produce more precise self-reports than those that are presented in the form of the client's other verbal and nonverbal forms of communication. Without the scale responses, the therapist must act primarily on the latter. The issue here has less to do with a contradiction than with the problem of securing reliable information to use as the basis for decision making. However, to answer the question, if the therapist has other information and on that basis is convinced the scale score is incorrect, the answer is obvious: do not depend on the scale.

If I see that a client's scores are increasing over a period of time, can I be sure that such an increase represents deterioration?

Again, yes, in nearly all cases. As with score declines, the increase must exceed the SEM for the scale and the amount of the increase nearly always indicates the amount of deterioration that has occurred. The ability of the scales to detect and report deterioration is one of their principal advantages because the presence of such deterioration alerts therapists to the need for strong, immediate, and definite action in most cases.

If a client's scores are steadily declining over a period of time, can I be sure that such progress is due to my treatment efforts?

Not unless you control for a number of other possible explanations. This is a complex issue that is described and dealt with in Chapter 5.

What should I do if a client does not want to complete the scales or refuses to do so?

Do not use the scales. However, by all means try to discover why the client refuses to cooperate. That information could provide important clues about the nature of the client's problem and the type of treatment that should be used.

Some clients will refuse to complete one or more of the scales, and in some of those instances the reasons may be related to the way the case has been managed. Refusals of this nature could mean that the client still does not have a good understanding of why the information is needed or how it will be used. If this problem occurs with a number of clients, the therapist might see a consultant or supervisor and role-play the manner in which the therapist is attempting to administer the scales. Then the supervisor or consultant might be able to offer suggestions that would help clients become more cooperative.

What should I do if scores from the scales radically depart from my own judgements about the severity of the client's problem?

Check the procedures you are using to score the scales, and reread Chapter 2. During a consultation with an experienced therapist, I was told that the scores obtained from the scales simply did not make sense because they were too high for five different cases. When examining one of the client's scales, it was found that the therapist made the proper item-score reversals, summed the scores for all 25 items, and then failed to subtract the constant of 25.0 to obtain the correct total score. This had happened in every case about which the therapist was complaining.

If it is certain that the scoring procedures are correct and the client's scores are still at great variance from the therapist's judgements about the severity of the client's problem, it is usually wise to discuss this with the client. It may be discovered that the client does not know how to respond to such scales, and in such cases it may be necessary to abandon them. In other cases, it may be learned that the client is deliberately manipulating the scores to achieve some purpose, and in other cases it may be that something is going wrong in the treatment itself or in the client's life. If all efforts to understand or resolve such differences fail, seek special consultation. Such conflicts in judgements or information should not be ignored.

Should I have all of my clients complete all of the scales?

No. It is unlikely that all of the scales will ever be appropriate for any specific case. They must be used selectively, and excesses in this regard could actually harm rather than help the treatment process. It is doubtful that therapists should ever have a client complete more than four different scales at a time on a regular or periodic basis and in most cases it will be sufficient to focus on only one or two. There are exceptions to these general rules but they tend to be rare.

How do I decide which scales to use?

The principal guideline to use here is to select the scale that measures the severity of the problem that the therapist is trying to change—the specific problem that will or has become the primary focus of treatment. However, one or two other scales might be used in some cases to monitor a closely related problem. For example, a therapist might work with a client whose major problem centers around disharmony in the marital relationship. The IMS scale would serve as the primary measure of choice in order to assess the client's problem and to monitor the course of treatment (see Chapter 5 for details). However, if the client is also depressed and has a problem in the sexual component of the spousal relationship, it may be wise to have the GCS and ISS scales completed as well. In such a case, the therapist might presume that progress in dealing with the marital problem could have a direct effect in terms of reducing both the depression and the sexual relationship problem. The evidence obtained from the ISS and GCS scales may be very useful in confirming or refuting such a presumption, but the major focus of attention would be on changes in the IMS scores over time.

Could I use a larger number of the scales during the intake and beginning phase of treatment as aids to diagnosis and case planning?

Definitely. From time to time we have all seen cases in which the client complains of many different personal and interpersonal relationship problems. For example, a client may complain of a marital and sexual problem, a problem with a specific child, depression, and poor self-esteem. In such a case it might be desirable to have the client complete the GCS, ISE, IMS, ISS, and IPA scales at the same time. These may be helpful in examining the severity of the problems in each area, and the findings may be useful in helping the therapist to decide which problems to focus on initially.

During the diagnosis and planning phase of a case, the therapist should use any of the scales or other assessment devices that are likely to help develop a sound treatment plan. One type of broad-based assessment device has been described in Appendix B and it might be very helpful if due regard is given to its inherent limitations. But, once the therapist has devised a formal treatment plan, subsequent measurement focus should be on a much smaller number of scales to effectively monitor the impact of treatment.

How often should I ask the client to complete the scales?

As often as the information is needed. However, it is generally good practice to have the scales completed on a *regular* basis. After all, one of their most important benefits is to give the therapist regular feedback about the presence and rate of progress or deterioration during treatment. If a short-term treatment plan is developed in which treatment will terminate after, say, eight to ten weeks, the scales should be completed once each week. If treatment is expected to continue for several months, they should be completed once every two weeks. Should treatment run beyond six months, the scales should be completed once every three weeks or once a month. The major consideration is to have the scales completed often enough to keep the therapist abreast of significant changes but not so frequently that they become an intrusion on the client or the treatment process.

Must the client always complete the scales in my office?

No. In most cases it may be wise to have the client complete one or two scales in the therapist's office so they can be scored immediately and the therapist can thereby benefit from the information. This might possibly help some clients understand how to answer the scales properly. However, the therapist may later have the clients complete the scales as homework assignments and then bring the completed scales to the next therapy session.

During a consultation, the author observed a therapist who made an appointment with a walk-in client. The client said he wanted to talk with the therapist about some problems he and his wife were having in their marriage. The therapist said, "All right, Jim, I can see you Thursday. I want you to take these three scales home (two each of the GCS, ISS, and IMS) and I want you and your wife to complete them and bring them back when I see you Thursday." Both willingly did so. The point is that the scales should be used flexibly and in a manner that is comfortable for both the client and the therapist. In many cases clients are *eager* to comply with reasonable requests.

Should I have clients complete the scales before or after each treatment session?

Again, flexibility and creativity are to be encouraged. However, it is probably best to have the scales completed before the treatment session begins. If they are completed at the end of the session, there is a risk that the scores will tend to reflect the emotionality of the specific treatment session; such responses could distort the total scores to some extent.

To what extent do the scales measure temporary or highly transient mood swings that occur over short periods of time and thereby disturb accurate problem assessment?

There is always some of that and it can never be totally eliminated. No doubt, such mood swings are part of the inherent noise or measurement error of each of the scales. However, the scales were not designed to primarily reflect such transient changes and experience to date indicates this is a minor problem.

It is important to distinguish between a transient mood swing and a substantial but short-lived affect change. For example, we may arise one morning feeling great, mellow down after breakfast, feel a bit reluctant to tackle a bad job at the office, become eager for an important business luncheon, feel grumpy late in the day, and then feel our spirits lift over a good meal in the evening with friends. The scales appear to be relatively insensitive to such daily swings of affect and problem encounters. Substantial changes can, however, be easily detected. A writer who was helping with some early testing of the scales almost always obtained a GCS score in the low 20s. One day he received a letter informing him that an important article had been rejected by a leading journal. When he thereupon completed the GCS his score was 42. His depression and dejection were very real and that was reflected by his GCS score. The next day his GCS score was again in the low 20s.

Should I allow clients to see their scores?

Definitely. Not only should clients be allowed to see their scores but therapists should help them to understand what they mean and how they will be used. If clients are not allowed to see their scores or helped to understand what they mean, the therapist may unknowingly imply to clients that the score outcome was devastating; that clients are too stupid to understand what all this means; or that therapy is an elitist system in which clients are not full partners in the efforts to solve their problems. Because some types of measurement devices have been too often used improperly (projective tests, IQ tests, and personality inventories) many of us have become a bit paranoid about the use of tests and scales. Secrecy concerning the clients' scores can stimulate suspiciousness on their part, and it also tends to inject an element of mystery into the treatment relationship that certainly is not needed and which rarely, if ever, is helpful.

Clients are not stupid; they have every right to know their scores. They are usually capable of understanding what the scores mean when properly explained, and they often make good use of the information. By showing clients their scores on the CMP scales and by helping them to understand what they mean and how they will be used, the therapist demonstrates respect for clients as intelligent people and shows that they will be treated as equal partners in the treatment process. In many cases, this procedure has an obvious effect of stimulating cooperation and enthusiasm on the part of the client.

Do clients show resentment or reluctance in completing the scales?

This seldom happens, provided the therapist does a good job of explaining why the scales should be completed, how they are to be used, and what the scores mean. Many clients are eager to learn more about the problems for which they are seeking help and that are making them miserable. As indicated earlier, one of the most important uses of the scales is in monitoring the client's problem over time, and when the scales are used in that manner, clients are often eager to complete the scales. In several instances, clients have asked if it is time to complete the scales again. This happens most often when real progress has occurred and the declining scores provide a distinct source of reward and an incentive to continue working on the problem.

In spite of the therapist's best efforts, however, some clients will display resentment, reluctance, anxiety, or other types of discomfort when asked to complete the scales. When that happens, therapists should always check out their own behavior and actions in using the scales to determine whether they are doing something to stimulate or prompt such reactions. If so, they should change their procedures accordingly. However, if therapists are confident their behavior is not responsible for such adverse reactions, those client reactions to the scales become important sources of information that can be used to help the client deal more productively with the problems where they are seeking help.

How should I interpret the scale scores to clients?

Nearly always with a great deal of candor. The first step is to explain to the client that the scales measure the severity or magnitude of the problem being examined—a large score represents a more serious problem than does a small score. While the therapist should be candid in interpreting the scores, an effort should also be made to avoid stimulating undue anxiety about the results. It goes without saying that therapists should not compute the client's score and then whistle through their teeth with surprise and alarm. It nearly always turns out that the scores obtained from the scales do little more than confirm the magnitude of the problem that the client is trying to convey through verbal descriptions and discussions. In those instances, it is usually sufficient to simply provide that confirmation. For example, if a client obtains a score between 30 and 50, the therapist might say something like, "Mr. Johnson, your score shows that you have a problem in this area, and though it's not terribly serious, it confirms what you've been telling me." If the score is, between 50 and 70, it might be appropriate to comment, "Mr. Johnson, this is a fairly large score, and it pretty well confirms all that you've been telling me about this problem."

Occasionally, scores will be obtained from the scales that do not appear to be congruent with the client's description or discussion of the problem. This must always be investigated to determine the source of the incongruence. It can be due to only the following: the client's discussion or description exaggerates or minimizes the severity of the problem; the scale is not valid for the particular client; the client did not understand how to complete the scale; or the therapist failed to properly understand the client's description and discussion of the problem.

What do very large scores of 70 or more mean, and what should I do when I see them?

Any time a score this large is encountered, it should alert the therapist to the possibility of violence. There is the distinct possibility that the clients may attempt violence against themselves or others. This is especially true for the GCS, ISE, CAM, CAF, IPA, and IPR scales. For the GCS and ISE scales, a score of 70 or more should cause the therapist to be on guard about, or to consider the possibility of, an attempted suicide. For the CAM, CAF, and IPA scales, scores of 70 or more should alert the therapist to the potential for child abuse or abuse of a parent, while for the IPR the threat of violence is toward one or more members of the defined reference group. For the IMS and ISS scales, scores of 70 or more could point to the potential for spouse or partner abuse and often mean that a divorce or separation is a distinct possibility.

Do not misunderstand. The CMP scales were not designed to predict suicide, violence against others, or divorce, and they should definitely not be used for such purposes. The foregoing is intended to mean that a score of 70 or more represents a very serious problem in the area being measured and that when people are caught up in personal and interpersonal problems they consider to be grave or catastrophic, they are often capable of a violent reaction or response. When clients obtain scores of 70 or more on the CMP scales, the therapist must not automatically conclude that violence is in the offing. However, it should be considered as a possibility and the possibility should be investigated in order to rule it out.

What if a client fails to complete a large number of items on a particular scale?

In such cases it is doubtful that the scale will produce a valid measure of the problem, and the therapist should not rely on the obtained score. Clients, on rare occasions, will leave a few items blank because they do not understand them or overlook them. When that happens, there is no great problem and the scales produce valid scores. In this regard, it is recommended that therapists do not rely on any scale score if the client fails to complete more than five items. But, remember, if any items are omitted, care must be taken to use the proper scoring formula as described in Chapter 2.

When clients fail to complete six or more items, that probably means the client has a literacy problem or that there is some special reason for omitting items. Most often it means the client has some special sensitivity to the item content or is resistent in facing up to and dealing with the problem. Although a scale score that is computed with a large number of item omissions should not be relied upon, such scale responses can be very valuable and should be used as a diagnostic aid. They provide important clues to resistence and the therapist should examine, with the client, the specific reasons for the refusal to answer the omitted items.

If I need two alternate form measures of one (or more) of the problems represented by the CMP scales, can I create such measures by splitting the scales?

This is definitely not recommended. The scales were not designed for this purpose, and there is some risk that they will not perform well if they are divided up to create alternate forms. This issue is discussed in detail in Chapter 7.

What if the client obtains exactly the same score on two or more occasions?

This can happen on two adjacent occasions but it is not likely; be on guard. It is very unlikely on three or more adjacent occasions, and in such cases the client may be giving social desirability or malingering responses aided by a superb memory. Check it out.

What do very low scores on the CMP scales mean: scores that range from, say, 0 to 5?

Assuming an accurate response to the scales, they indicate that the client has no problem in the area being measured and the therapist should not be treating such problems.

Does a score of zero on the CMP scales mean that the person has none of the problem?

Strictly speaking, from a measurement theory point of view, the answer is "No." For any conceivable *practical* viewpoint or application the answer is definitely, "Yes." For all practical applications, the CMP scales can be treated as true ratio scales that are assumed to have a natural origin and equal measurement intervals. These assertions do not hold up under close scrutiny, but in practice it does not seem to make any difference as to whether these assertions are strictly true or strictly false. More research is needed in this area.

How do cultural differences affect scale scores and may I use the scales with clients from different ethnic and cultural backgrounds?

More data are needed in this area, but the evidence to date, for example, Murphy, Hudson, and Cheung, (1980) suggests that cultural and ethnic differences have little, if any, impact on the performance of the scales. When using the scales with clients from markedly different cultures, the major factor to be confirmed is that the client has a good understanding of the item contents and their meaning. At some point this becomes a problem of language and if English literacy is a problem, the scales must be translated into the client's native language. The scales have already been translated into Chinese, French, German, and Spanish, and they appear to perform well in those forms.

Can I use the scales with mentally retarded or mentally handicapped people?

The scales should never be used with persons who are severely mentally retarded. However, they might certainly be used with some persons who are victims of certain types of brain trauma and with persons who are mildly retarded. For example, the scales can almost always be used with persons who have cerebral palsy. Again, the critical question is one of whether the client can understand and properly respond to the item content of the scales.

Can I use the scales with psychotics?

Not when the psychosis is in an acute phase. However, therapists should not presume that the scales cannot be used with clients who have a diagnosis of psychosis. If the acute symptoms are in remission, the scales may be very useful. Therapists should use (or not use) the scales in accord with their best judgements concerning the utility and appropriateness of the scales and in accord with how the stability of the client is assessed.

Can I use the scales with involuntary clients and people who are incarcerated?

The CMP scales should not be used with any involuntary client unless and until the therapist has established good rapport and *mutual* trust. Incarcerated people are a very special class unto themselves. Any person who has been given an indeterminant sentence that can be shortened as a function of demonstrated good behavior is not a likely candidate for providing candid responses to the scales. *Proceed with caution.* The score results that are obtained from such clients may be totally misleading.

It appears that these scales may be very useful to me and some of my coworkers in our agency, but we already have so much paperwork that I am not sure we could ever use them. How do you suggest we deal with that?

Many agencies, unfortunately, have excessive amounts of paperwork and in some agencies it is oppressive. Not all gets done, and therapists are often in a position to decide what priorities to give to what types of record keeping and paper processing. If these scales are used to monitor and evaluate the work therapists' do with clients, there is no question but that such usage will add to existing paperwork burdens. However, some paperwork adds greatly to therapists' performances in helping their clients and is therefore essential. Some adds very little to such performance and, frankly, therefore needs to be reduced or controlled. If it is felt that use of these scales will be helpful, it might be worthwhile to try and negotiate paperwork priorities with coworkers, supervisors, and administrators within the agency.

If therapists decide to use these scales, they will discover that they will be intrusive initially; the introduction of any new paperwork system is nearly always intrusive. However, it will also be discovered that the use of the scales requires less time than might be thought at first, and after therapists gain experience with the use of the scales, they will likely be surprised how small the imposition that is actually created. Most important, their utility is usually so great that this nearly always compensates for time lost in their initial use and application.

Can I require clients to complete the scales as a condition of treatment?

Yes, but this should very definitely be a *negotiated* rather than an *imposed* requirement. Therapists should give clients a clear and complete explanation of why they are using the scales and how they will be used to help the treatment process. Some therapists feel very strongly that if there can be no mechanism for monitoring and evaluating treatment, there is some basis for questioning the legitimacy of engaging in a treatment relationship.

Can I use the scales in supervision and consultation?

Definitely. They are very often quite powerful aids to the supervisor and the worker, and this is true especially when the scales are used in accord with the material discussed in Chapter 5. A simple chart or graph of the scale scores over time, often with annotations posted on the chart, can be a powerful device for efficiently summarizing an entire case history on a single page. Such devices are enormously helpful to consultants who must quickly understand and respond to a large amount of information concerning a single case situation. Some supervisors have found the scales and the materials presented in Chapter 5 to be so helpful in the supervisory role that they have taken the initiative in requesting that their workers use these and similar materials.

Can I use the scales as teaching aids in working with students?

Yes, indeed, both in the classroom and in practicum training. The CMP scales, other measurement devices, and the materials described in Chapter 5 have frequently been used as potent training tools. Students are usually very enthusiastic when they learn that they can obtain improved evidence of client progress or deterioration. When case materials are presented for discussion, it is nearly always found that case planning and treatment issues are more sharply focused and identified when evidence from the use of the scales is presented. The scales themselves and the charts and graphs discussed in Chapter 5 are highly useful as handouts and as the basis for slide and overhead projector presentations. Some instructors will use this field manual extensively as a supplementary text for a variety of treatment or practice courses; some practicum instructors and supervisors will use it routinely as a training and teaching aid.

What are some of your suggestions for using these scales in classroom and practicum training settings?

There are several ways this can be done. First, this manual can be used as a supplemental text in the classroom. It is especially useful in this way because it provides a basis for training students to use formal tools of measurement as an aid in assessing the severity of client problems. It also introduces the technology of single-subject designs as a strategy for monitoring and assessing client progress and the impact of intervention. The use of this manual is strongly encouraged as a required text in at least one course devoted to clinical practice in which all students will enroll.

The manual is also especially useful in any idiographic research methods course that is designed for clinicians. In such a research course, and others too, the manual provides an excellent basis for introducing and supporting lecture material devoted to assessment and measurement theory and methods; the scales provide concrete and highly useful working examples of applied measurements.

In practice courses and practicum training settings, students should definitely be encouraged to complete the training exercises in Appendix A. Also, classroom instructors may want to work with field practicum or internship instructors to acquaint them with the manual and to help provide students with opportunities to use all of these materials in their practicum or internship settings. Many homework exercises can be designed around the use of the scales, but in the final analysis the best training comes from having students use them in practice when working with clients.

Should I complete the scales myself?

Definitely. Therapists who plan to use the CMP scales with clients should complete all of the scales that apply to their own personal situation. Moreover, they should complete the scales once a week for a period of four to six weeks as part of their own training in learning how to score and interpret the scales. Their results should also be graphed in order to learn how stable or variable their own responses will be (see Chapter 5 and Appendix A). The knowledge that is gained from such experience will help therapists to use the scales more effectively with clients and to understand better the scores provided by clients. There are several self-administered training exercises included in Appendix A and all of those exercises that are appropriate to the therapist's personal situation should be completed.

How can I protect against "impression management" types of responding?

The best way to protect against or minimize social desirability, malingering, and other types of "impression management" responding is for therapists to convince the client that the therapist has a legitimate reason for obtaining accurate responses to the scales and that such data will be useful and helpful to the client. Much of this depends upon how well the purpose and use of the scales are explained to the clients and how well the therapist explains the monitoring and assessment functions of treatment. As might be expected, it greatly helps to develop and maintain rapport in the treatment relationship.

In spite of the fact that all of the CMP scales are obvious in their measurement intent, evidence to date suggests that respondents provide candid and accurate responses to all of the scales when they are used in a legitimate manner and under proper conditions. However, it should be remembered that "legitimacy" and "propriety" must be defined in terms of the client's viewpoint and not the therapist's. If the therapist believes that some use of the scales is legitimate but the client does not, the client may think the therapist does not have a right to the information, and, in such cases, it is not likely to be forthcoming.

Do clients ever falsify responses to the scales in order to delay or speed up termination of treatment?

Of course, some do, but these are relatively rare events. When they occur, attention should not be focused on the fallibility of the scales but on the treatment process and its management. If the client's problem has been solved and their is no need to continue treatment, the therapist should investigate carefully why the client might want to continue a treatment relationship. If the client's problem has not been significantly reduced or solved, the therapist should investigate why the client might want to terminate treatment prematurely.

Should I use a "social desirability" or "lie" scale along with the CMP scales?

No. Such scales might show that the client is lying or giving socially desirable responses. They might show that the therapist is not getting truthful or accurate answers but they certainly will not tell what the truth might be. Worse yet, their use may imply to clients that the therapist does not trust them and are unwilling to do so. Keep in mind that if the therapist explains the use and purpose of the CMP scales, what is the client to be told about the use and purpose of a social desirability or lie scale? The client is likely to ask. The best way to handle this problem is to simply cross-check the scale scores with all other data that are available about the client's problem.

If I use any of the CMP scales in a research project, can I assume the scales are normally distributed?

No. The CMP scale scores are not normally distributed in the general population and they are not expected to be so distributed. At any point in time most people do not have a clinically significant problem in any of the areas represented by the CMP scales, so most people obtain scores that fall below 30. This means that all of the scales have score distributions that are moderately to severely right-skewed. This generally poses no problem because the skew that is seen in a representative sample generally does a good job of describing the skew that is present in its parent population. The presence of such skew can, however, affect the accuracy of statistical tests that are based on the normal curve. For improved accuracy in hypothesis testing, it is recommended that one of the tail-shrinking transformations be considered. Generally speaking, a square root, logarithmic, or logit transformation will suffice.

Can I factor analyze the CMP scale items for any specific scale in order to determine what types of problems clients are having in each of the areas being measured?

No. The CMP scales were not designed as multidimensional scales and they should not be used as such. Chapter 7 has a more detailed discussion of the factorial validity of the scales.

If I decide to use the CMP scales as training aids in the classroom and in practicum settings, when should this be done? At what stage in the training program?

As soon as possible. Students should begin in the first or second semester (or quarter) of their academic program to learn the fundamentals of monitoring and evaluating their clients' problems. The sooner they become exposed to such technology, the sooner will they be able to assess better their own ability to function as effective therapists in dealing with simple and complex client problems. Knowledge that is gained early in their program of study will serve as a foundation upon which to develop further expertise in their treatment and assessment technology. Although this field manual is focused on the use of the CMP scales, students should be trained in the proper use of a large number of other measurement tools.

REMARKS

As others gain more experience in using one or more of the CMP scales, many additional questions are likely to be raised. This chapter does not intend to treat all of them. However, as new questions arise, the author would like to have them. If they cannot be answered at this time, solutions may be found from new research and from extended clinical usage.

5 Monitoring and evaluating clinical practice

The CMP scales were developed and validated primarily for use in monitoring and evaluating clinical practice. This chapter provides a brief description of simple but highly effective methods that can be used to monitor client progress in treatment and to shed some light on the effectiveness of treatment methods used by the therapist. By using the methods and procedures described in this chapter, it is likely that the therapist's ability to understand the client's problems will increase and that the therapist will become a more effective professional helper.

WHY EVALUATE PRACTICE?

There are many good reasons for evaluating one's practice to determine whether the intervention methods that are being used are effective in helping clients to solve, reduce, or eliminate problems. Most of those reasons are sound and valid, but some are misdirected. For example, it might be claimed that clinicians should evaluate their professional intervention to demonstrate that they are "good" therapists. Actually, that is an unimportant objective of evaluation. Clients want their therapist to be regarded as good therapist, but only because that will mean there is a better chance that the therapist will be successful in helping them to solve their problems. The rest of it, the ego trip, is an act of self-indulgence. If therapists wish to pursue it, that will be their decision, but from a professional point of view, there is little to be gained in showing that one is a great or a lousy practitioner.

Another reason for evaluating practice arises from the need to be accountable. We live and work in an age of increased accountability, and it appears that such an emphasis will continue or perhaps grow more intense as service needs continue to increase and professional resources become scarcer. Funding agents, policymakers, and upper- and middle-level managers and administrators must increasingly seek and present information to show that a huge variety of human service programs are doing the job for which they were designed. However, from the practitioner's point of view, an emphasis on accountability means that the clinician must respond to organizational and institutional agents rather than to the client who is seeking help.

This perspective in no way suggests that the need to show that therapy is effective is an unimportant task. Entirely too much evidence has been presented to raise serious doubts about whether we help anyone. It is my conviction that there is some merit in this evidence but that the case is too often overstated. The truth will probably be found somewhere between one extreme that claims therapy to be almost always useful and another that claims the opposite. To avoid complacency in this matter, we should all be thoroughly familiar with at least some of the research evidence concerning prior attempts to evaluate the effectiveness of clinical and other forms of human service efforts. This manual does not represent an appropriate forum for discussing such studies, but the interested reader may wish to consult a few useful references (Eysenck, 1952; Mullen and Dumpson, 1972; Fischer, 1973, 1976; Garfield and Bergin, 1978; Wood, 1978). An excellent brief review of such material can be found in the text by Zastrow (1981).

A third major reason for evaluating practice is one that emerges from the domain of science. In this context, the fundamental purpose of evaluation is to discover what types of treatment techniques will be effective in reducing or solving what types of problems with what types of clients under what types of conditions. The fundamental aim is to develop an empirically validated knowledge base that practicing professionals, classroom and practicum instructors, and students can draw upon to upgrade existing competencies and to acquire new ones. Herein lies the major task, function, goal, or purpose of research and scholarship in the helping professions. No doubt, some practicing clinicians and therapists do engage in such basic research and scholarship while conducting their major work as practicing professionals, but such persons are exceptions. Most practicing therapists do not engage in such basic research and for them an emphasis on the knowledge-building and theory-testing functions of evaluation is one that also detracts from the primary goals and purposes of professional practice.

In short, most of the reasons for conducting evaluations of practice are seen by most clinicians to be an intrusion upon practice and not as an aid to support it. Fortunately, there is a fourth reason that can be advanced as a justification for evaluating clinical practice, and it is the one that is made predominant here. According to this position, *the most important purpose for evaluating clinical practice is to acquire information that will help the therapist to increase the likelihood of a positive outcome for the client* (Hudson, 1978b). In subscribing to this view, it should be said that the fundamental aim of evaluation is to help the therapist and the client, and not to satisfy accountability needs or to produce new knowledge for science. As it turns out, information that is gathered primarily for use by therapists and clients often goes a long way toward meeting accountability obligations, and such information often does a credible job of furthering the aims of science. Nonetheless, those are seen here to be subsidiary rather than primary goals. Here, the basic techniques of assessment and evaluation will be described for only one purpose: to help the therapist to improve or increase the likelihood of a positive outcome for the client.

A BASIC EVALUATION STRATEGY

The basic strategy behind the evaluation of clinical practice is incredibly simple. It consists of simply measuring the client's problem repeatedly over time and then using the obtained problem-score information to make judgements about the case. That's all there is to it!

Suppose, for example, that Mrs. J., a 57-year-old woman, was referred to a counselor because she was depressed. The intake counselor believed that Mrs. J.'s depression was largely due to the death of her husband three weeks earlier and that the fundamental task was to help her adjust to the loss of her spouse and to a new lifestyle as a single person. On the basis of the information that is contained in the intake interview summary, the therapist decided the only thing that was needed was that the client have someone with whom she could talk as she worked through her sense of loss and adjusted to living alone.

It was the therapist's opinion that the client's depression would decline if the therapist would simply listen to her problems and her grief, offer sympathy and understanding, and give advice, when appropriate, about such matters as money management, shopping, new sources of leisure and recreation, and so forth. The therapist decided to see the client once each week. To check on her progress in treatment, she was asked to complete the GCS scale at the beginning of each treatment interview. The GCS scores that were obtained for this client are shown here.

Treatment session	GCS scores
1	64
2	61
3	65
4	58
5	61
6	63
7	59

Without knowing any further details about this case, the therapist amassed a great deal of important information. The first GCS score showed that Mrs. J. was very depressed, and the pattern of scores over the seven-week period of treatment showed rather dramatically that Mrs. J.'s depression did not abate. What is the therapist to do? Is it the therapist's judgement that the choice of treatment is a good one? How soon can the GCS scores be expected to decline? If the therapist now thinks the treatment should be changed, what are the kinds of changes to be made? If the nature of the treatment is changed, how soon can one expect the GCS scores to decline?

INFORMATION AS FEEDBACK

The case of Mrs. J. was introduced to illustrate and dramatize the single most important use of the CMP scale scores in monitoring and evaluating practice: feedback to the therapist. The use of the CMP scale scores as a feedback mechanism is so powerful that it is difficult to exaggerate their importance in this regard. If used as a feedback mechanism, the CMP scales and many other types of measurement devices can come very close to transforming all types of therapy or direct service into a *self-regulating process* that has the central goal or purpose of helping the client.

When one engages a client in therapy or undertakes any professional commitment, there must always be a planned expectation that one's service or treatment strategy will be effective and that, as a consequence, the client will experience some benefit from the service that is provided. There are two basic issues here: whether treatment is effective and whether the client benefits. From a scientific point of view, major interest is always focused on the former, but from a practice point of view, the major emphasis is always on client improvement. Evidence concerning the effectiveness of treatment is *always* deduced from evidence concerning client benefit or improvement.

This paradigm lays the groundwork for defining the concept of feedback in clinical practice and then using it to help clients. Thus, after beginning a specific treatment program, each time the client completes one of the CMP scales it is fully expected that the obtained score will produce some evidence of improvement. It is in this sense that the CMP scale scores provide feedback to therapists about expectations of positive changes; the obtained scores provide feedback information about the extent to which expected change did or did not occur. If therapists will use such information about the presence of absence of desired change as a basis for making decisions about continuing or modifying treatment methods, their professional practice will function largely as a self-regulating mechanism that adjusts itself to increase the likelihood of a more positive outcome for the client.

To avoid misunderstanding, it must be said that positive change cannot be expected to occur every time a client completes one of the CMP scales or any other measurement device. However, a therapist cannot reasonably allow a program of treatment to remain unaltered if long periods of time and many repeated measurements of the client's problem produce no evidence of growth. The major point is that it is the *pattern* of CMP scores over time that provides the essential feedback information about the presence or absence of client growth, stagnation, or deterioration.

CHARTS AND GRAPHS

As stated above, the basic strategy that is used to monitor and evaluate clinical practice consists of simply measuring the magnitude of the client's problem repeatedly over time and then using the CMP scale scores as a basis for making judgements about the presence or absence of client growth and about retaining or altering the nature of treatment. When using this simple evaluation strategy, there are a number of interpretational aids that should be considered because they often help enormously to better understand the pattern of scores that is obtained from use of the CMP scales. One of those aids is a chart or graph of the CMP scale scores over time. This simple device is considered so helpful that it is recommended for use with all cases when appropriate.

In preparing charts and graphs, it is strongly urged that therapists develop a standard form and then reproduce a large supply of blanks. Therapists can design and then reproduce their own blank charts and graphs, or specially prepared blanks can be ordered from the publisher. The writer has found that the blank chart shown as Figure 5-1 is flexible enough for use in most situations, but readers may wish to experiment with a variety of different forms designs.

In order to prepare a chart for a client, the therapist needs only to fill in the information shown at the top of the chart and then plot the client's scores from one or more of the CMP scales. Figure 5-2 is a chart that was prepared for Mrs. J., and in order to prepare it, the worker merely plotted the GCS score for each treatment session and then connected the plotted score points with a straight line. The fundamental advantage of a chart, such as the one shown as Figure 5-2, is that the reader can readily see whether and to what extent

FIGURE 5-1

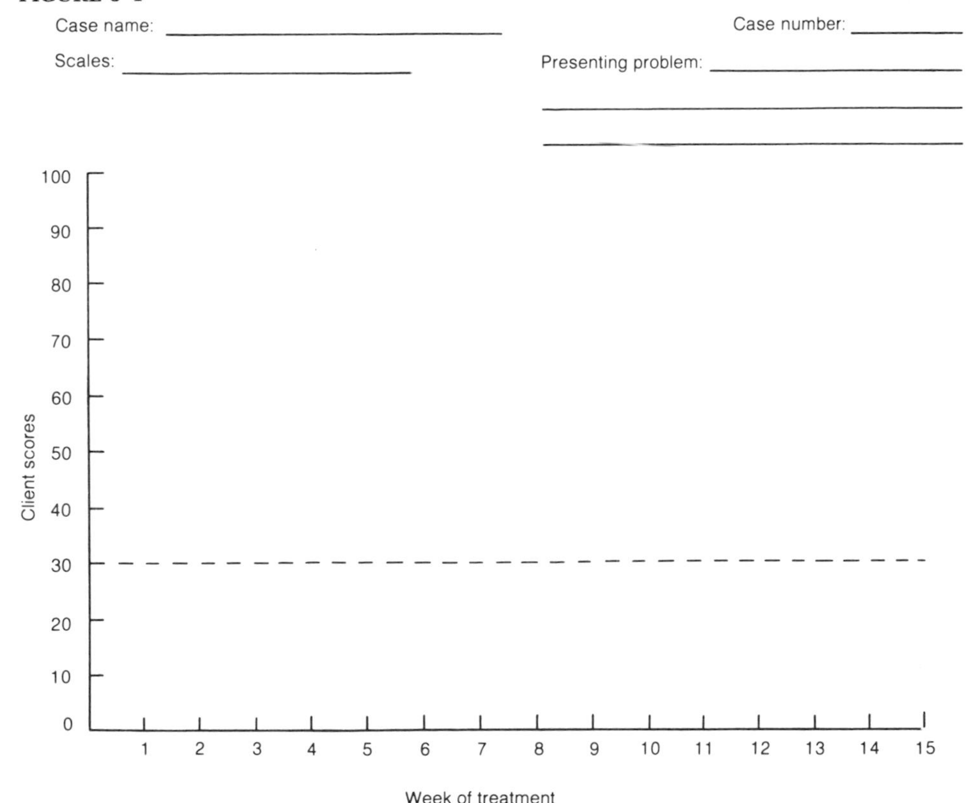

FIGURE 5-2

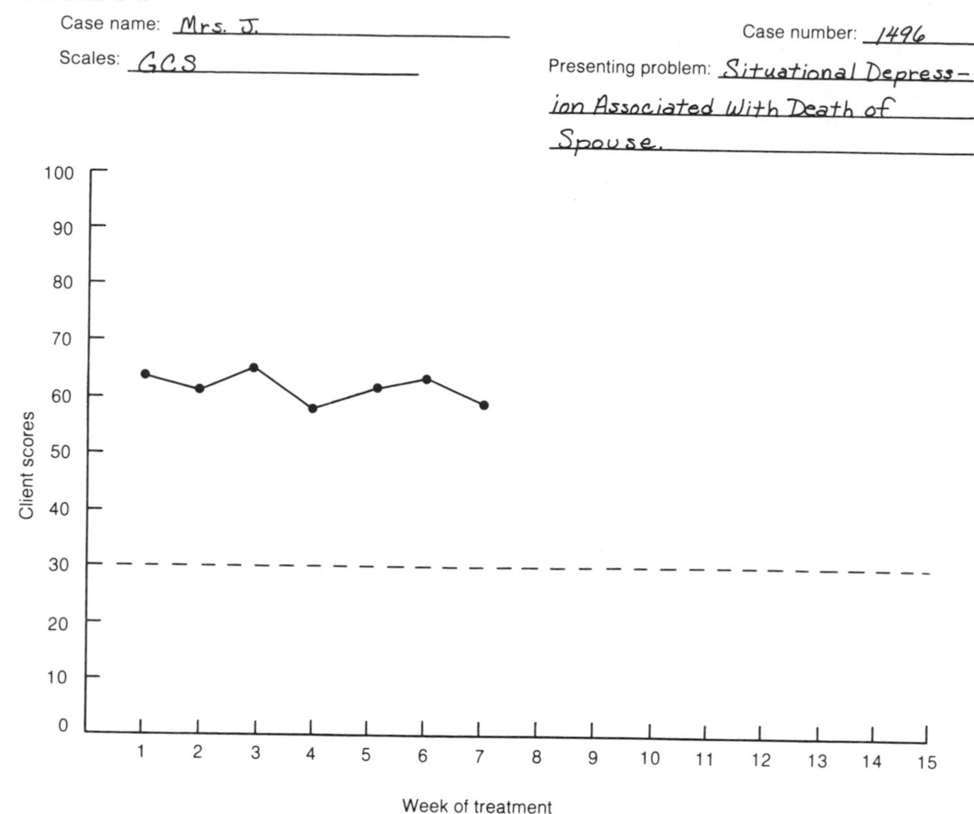

there has been any change in the magnitude of the client's problem. Moreover, the use of a well-prepared chart can serve as a powerful aid to improve communication with a supervisor or case consultant, and some supervisors require their workers to maintain such charts for all cases when that is appropriate. A chart such as the one shown for Mrs. J. provides the busy therapist with a very quick summary of the entire case, and it often serves as a reminder of important details that otherwise might lie buried in the case record.

It is not necessary for the therapist to prepare a separate chart for each different CMP scale that may be used for a particular case. Instead, it is often helpful to graph the results from two, three, or even four different scales onto a single chart. Figure 5-3 is an example of a chart that was prepared for a male client who was seeking help primarily for a problem he was having in his relationship with his wife. In addition to monitoring and evaluating this client's progress in dealing with marital relationship problem, the therapist also

FIGURE 5-3

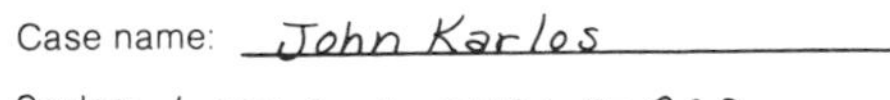

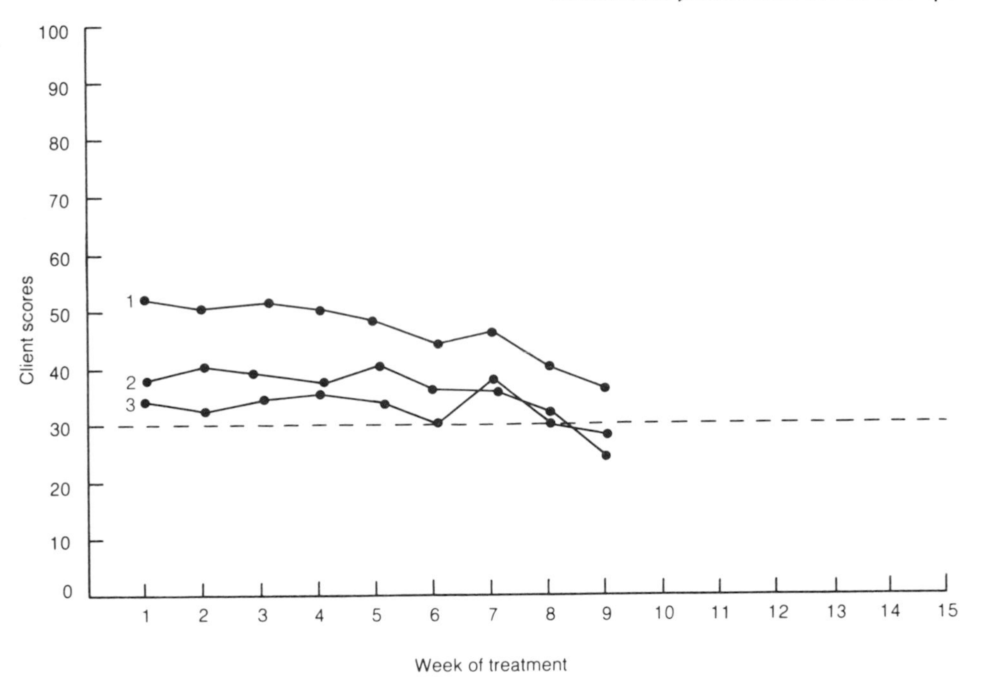

monitored the client's level of depression and the level of sexual discord in the spousal relationship. Because graph lines can become confused, it is not recommended that more than four graphs be placed on any single chart. In some cases, the therapist may want to prepare two separate charts to record the scores that are obtained from four different scales.

CRITICAL INCIDENT RECORDINGS

Another type of interpretational aid that should be kept for each case is referred to as a Critical Incident Recording or CIR. It is easy to use and consists of a brief description, by session and date, of those specific events within the treatment relationship and in the client's life which the clinician considers as having significance in relation to the problem that is being treated. Figure 5-4 presents a format for a critical incident recording sheet that I have found to be very useful.

FIGURE 5-4

Critical Incident Recording Form

Case Name: ______________________ Case Number: ______________

Week	Date	Comment

For those who have never used such devices as charts, graphs, and CIR forms, it is tempting to overlook their importance. The great power of the CIR arises from the fact that the therapist can associate the CIR entries with the CMP scores on a chart, and an analysis of these two types of information can provide important clues about the kinds of events and situations that may be responsible for change in the level or severity of the client's problem. Some events may be so potent that the therapist will want to record a note of their occurrence on the chart itself. However, such a practice should not cause therapists to abandon use of the CIR form.

When using the CIR form, the therapist may be tempted to make increasingly more elaborate case notes, and if that happens, the CIR form will lose its major utility. Clarity and brevity are hallmarks of good CIR, and the CIR form should be restricted to the recording of specific behaviors and events that the therapist thinks are, or may be, clinically significant. Diagnostic speculations and process notes should not be placed on the CIR form. An example of how one worker used the CIR form is shown in Figures 5-5 and 5-6. There it can be seen that several specific events the therapist thought might have been of some clinical significance apparently had no effect on

FIGURE 5–5

Case name: B. Tompson

Case number: 4T7063

Scales: 1 = GCS; 2 = ISE

Presenting problem: Acute Depression Associated With Change of Employment; Wife May Separate.

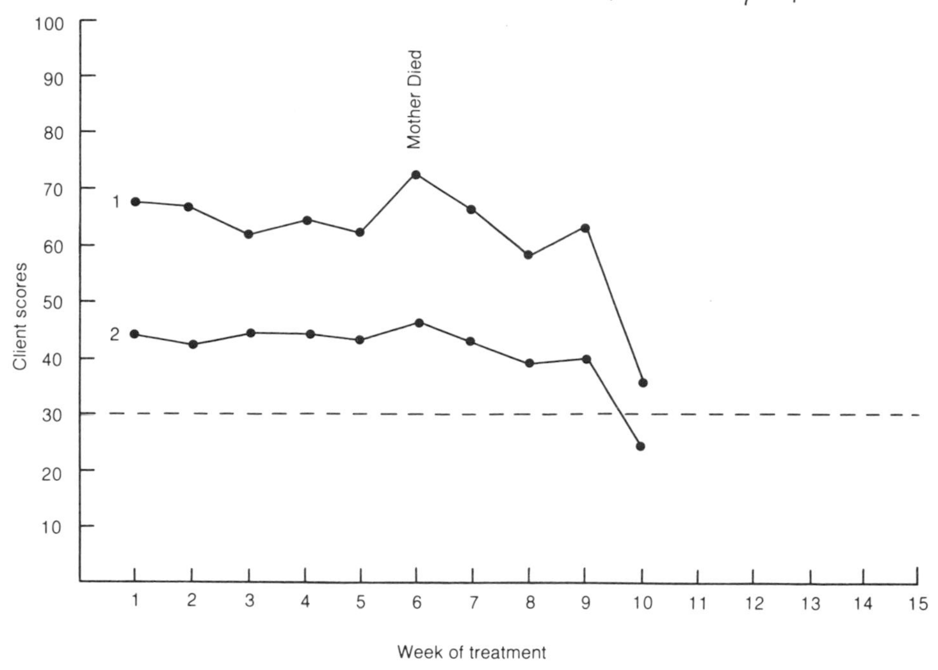

FIGURE 5–6

Critical Incident Recording

Case Name: B. Tompson

Case Number: 4T7062

Week	Date	Comment
1	04/04/78	Severe depression associated with loss of employment. Has no job prospects.
2	04/11/78	Mr. T. got appointment with State Employment counselor.
3	04/18/78	State Employment Office has no jobs in Mr. T.'s area.
6	05/09/78	Mr. T. spent hour talking about his mother's death of four days ago.
8	05/23/78	Mr. T. said 17-year-old son was arrested for driving while intoxicated.
10	06/06/78	Mr. T. landed a top job two days ago. Will earn higher salary than on previous job. He's a bit scared of the heavier job responsibility he'll assume.

the client's problem, but two events were associated with client change.

VALIDATING SITUATIONAL DISORDERS

In clinical practice, it is rather common to distinguish between situational disorders on one hand and chronic or psychogenic disorders on the other. In the case of Mrs. J., it was presumed that her depression was a situational disorder that arose as a consequence of her husband's death. In fact, Mrs. J. told the intake counselor that she became depressed "at the time of my husband's death" and she also told the therapist that Mr. J. had always been "a picture of health"; his heart attack happened with no warning. Armed with this kind of information, it may seem quite logical and reasonable to assume that Mrs. J.'s depression was a situational reaction.

Since Mrs. J. had been in treatment for seven weeks and there was no apparent decline in the level of her depression, the therapist decided that an attempt should be made to verify the presumption of a situationally induced depression. In order to do that, the therapist asked Mrs. J. to think back to the week immediately preceeding Mr. J.'s death and then fill out the GCS scale in terms of the way she had been feeling at that time. The therapist then asked Mrs. J. again to fill out the GCS but in terms of the way she felt about a month before Mr. J.'s death. The therapist scored the two GCS scales and then plotted them on a new chart, as shown in Figure 5–7, along with the scores that were obtained during the seven weeks of treatment.

Inspection of Figure 5–7 strongly suggests that Mrs. J.'s depression was *not* situationally induced as a function of the death of Mr. J. The data presented in Figure 5–7 indicate that Mrs. J.'s depression actually began at least a month prior to her husband's death, and this indicates that the intake counselor and therapist had developed an incorrect diagnosis and continued to work with it.

The important point of this example is the suggestion that it is extremely important to verify any diagnosis of a situational disorder. Ideally, the therapist would want to do that by collecting information about the client's problem before it reached a clinically significant magnitude. For obvious reasons, that is not possible. However, as in the case of Mrs. J., it may be quite easy in many cases for clients to complete one or more of the CMP scales in terms of the way they remember their thoughts and feelings prior to the event or situation that is presumed to be the cause of the current problem. If the diagnosis cannot be confirmed through the use of such retrospective information, it may very well be incorrect.

FIGURE 5-7

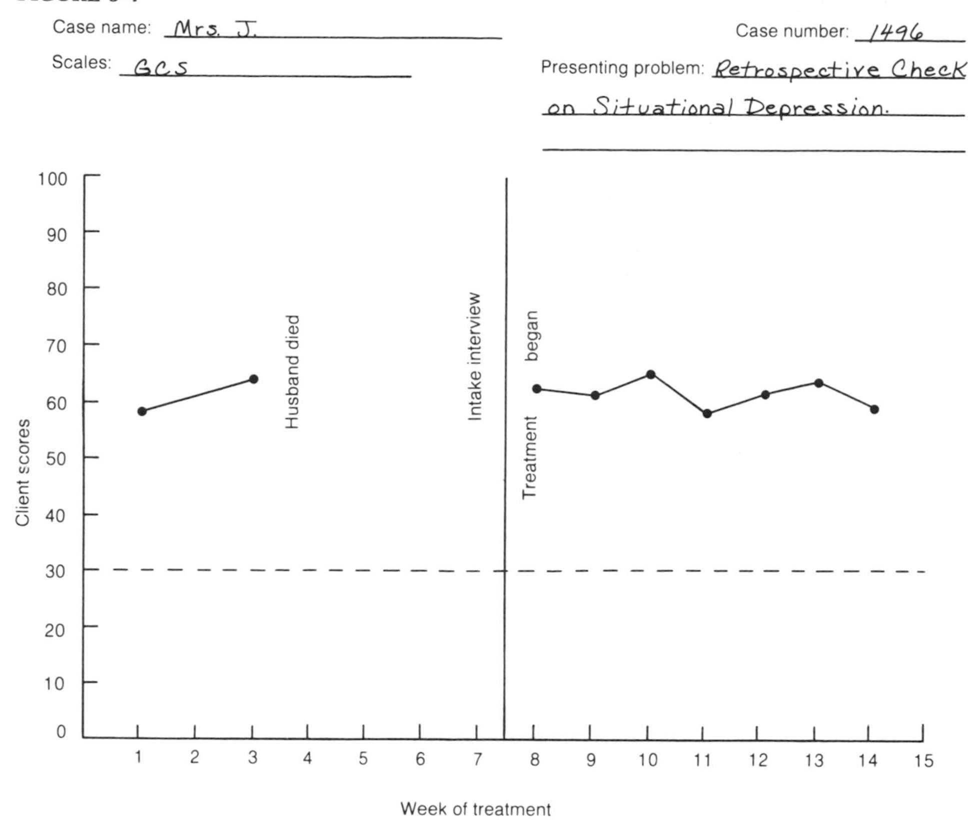

BASELINING

The concept of baselining is very important to clinicians who work with a wide variety of clients and with different types of problems. Baselining consists of collecting information about the magnitude or severity of the client's problem during a period prior to the onset of treatment. The data that are obtained during that period are referred to as the *baseline measurements* or, more simply, the *baseline*. In the previous section, the therapist asked Mrs. J. to complete the GCS scale in terms of the way she felt one week and one month prior to the death of her husband. Those two GCS scores would properly be referred to as baseline measurements. Because Mrs. J. was asked to provide those data in terms of how she best remembered her thoughts and feelings, those GCS scores are referred to as *retrospective* baseline measurements or as a *retrospective baseline*.

There are two very important reasons for collecting baseline measurements. First, baselining is nothing more than a somewhat more formal means of conducting an ordinary case assessment. Clinicians normally devote a significant amount of time and energy to the task of assessing client problems during the period preceeding the onset of formal treatment. Therapists must ask a range of questions about the origin, severity, onset, and current status of the client's problem to obtain ideas and clues about how best to treat it. Virtually all of that information can be regarded as a form of baselining, and because the information is obtained with respect to time periods preceeding the onset of formal treatment, it might be regarded as a form of retrospective baselining. In other words, baselining is an integral part of initial case assessment. One might even say that initial case assessment is merely the retrospective baselining phase of treatment. It must be done for each and every case in order to form a basis for developing a treatment plan.

The second important reason for collecting baseline information is to provide a basis for comparison with the intervention phase of treatment. It is most unfortunate that this notion of comparing baseline levels of the client's problem with the intervention phase levels of the problem has been most often associated in the scientific literature. In fact, this notion is, and always has been, a centrally important and fundamental concept of all forms of clinical practice: The basic purpose of therapy is to help clients "get better," but "get better" with respect to what? This means with respect to the level of the problem at the time they seek treatment and with respect to the level of the problem during some period preceeding the onset of treatment.

As indicated earlier, the fundamental aim of all professional practice is that the conduct of therapy produces some form of measurable or verifiable benefit for the client. In this context, the concept of *benefit* must be defined in terms of some type of baseline information or measurement. If, at some time during or *shortly* following the intervention phase of treatment, the client is not improved with respect to the level or severity of the problem that was observed during the baseline period, it is not possible to demonstrate that professional practice produced any benefit for the client. In short, client benefit is always defined and measured in terms of the difference between the baseline and intervention phases of treatment with respect to the level or severity of the problem that is being treated.

COLLECTING BASELINE DATA

Since baselining is central and crucial to the conduct of professional practice, it is very important to examine the basic strategies for obtaining such information. No doubt, the initial intake interview with a client is one of the single most important devices for collecting many different types of baseline information or data. The intake counselor must obtain a great deal of information about the client's complaint, its onset and duration, previous efforts to deal with it, and the individuals, groups, or institutions that are involved with the problem. In many cases, not all of this information is retained. The intake worker must think about and analyze the information that is obtained from the client and possibly collateral sources, and a major goal for both the intake counselor and therapist is to use all of the baseline information to construct a well-defined statement of the problem to be treated.

An important point here is that during the initial information gathering interviews most, if not all, of the baseline data that are obtained about the case are not stored or processed in *measured* forms. However, once the client and the therapist have reached some agreement as to the specific problem or problems that will become the target of treatment, it is then important for the therapist to consider ways of obtaining measured baseline assessments of the problem to be treated. The purpose for obtaining measured forms of the baseline data is to produce a more precise assessment or description of the severity of the problem and to use that as a basis for defining and evaluating the benefit to the client as a consequence of professional practice.

Although there are many different ways to obtain measured baseline assessments of clients' problems, a discussion of them would go beyond the purpose of this manual. However, if the client's problem is one that can be measured and assessed by using one or more of the CMP scales, it is important to consider using these scales to obtain baseline measures of the problem. If the initial intake counselor is in a position to reach closure in defining the problem to be treated, it may be possible to collect some baseline data by using the CMP scales during the intake interview.

If a counselor decides to use one or more of the CMP scales during the intake phase of treatment, it is best to use them to obtain current-status measures of the problem before using the scales retrospectively. For example, a therapist might say to a client, "Mr. B., I'm getting a good picture of the problem you are having with your son, but I want you to fill out a short scale (the IPA) that will tell us both more about how serious you really think this problem is." After the therapist describes the scale and the client completes it, the therapist might say, "Mr. B., I'd now like for us to get a better picture of how serious this problem was when John dropped out of school. You said that was about the time when you and he had the greatest problem in working and getting along together. I'd like for you to think back to that time and fill out the IPA scale again in terms of the way you felt when John dropped out of school."

At this point, the therapist has obtained a measured assessment of the severity of the parent-child relationship problem, as seen by Mr. B., in terms of his current view of it and in terms of how he saw it at its most severe point. The therapist might consider asking Mr. B. to complete the IPA two more times: once for the period of about a week preceeding John's withdrawal from school and once for the period of about a week preceeding Mr. B.'s appointment with the intake counselor. If that were done, the therapist would have four important baseline measures of the severity of the parent-child relationship problem as seen by the client.

Some therapists may feel that it is not wise to ask a client to complete one of the CMP scales as many as four times during a single intake interview, and one certainly must be sensitive to the possibility of overkill in collecting baseline data of this type. Nonetheless, in a surprising number of cases it is entirely possible to use one or more of the CMP scales in a manner such as described above. Much of the success in doing that depends on the therapist's skill in demonstrating to the client a need for such information and in giving clear and simple explanations of how that information will be used by the client and the therapist to monitor and to evaluate progress during treatment. Also, if it is the therapist's judgement that one or more of the CMP scales should not be completed one or more times during the intake interview, it may still be possible for the intake counselor or the therapist to obtain such baseline information during subsequent interviews.

CONCURRENT BASELINES

Concurrent baselines are defined as those measurements of the client's problem that are obtained during the period beginning with the intake interview or phone contact and running through the period preceeding the onset of a formal intervention strategy. In the case of Mr. B., the client was referred to and seen twice by a therapist before the therapist began a formal program of intervention. In this case, the therapist used a form of role-playing to help Mr. B. get a better picture of his son's point of view and to help Mr. B. better understand how he was overreacting to his son's problems in school. Figure 5-8 provides a description of the entire case, and it is shown here primarily to illustrate the use and definition of three distinct phases of treatment: retrospective baseline, concurrent baseline, and intervention.

Some clinicians have a great deal of trouble with the use of baselining, and this occurs, in part, on grounds of professional

FIGURE 5-8

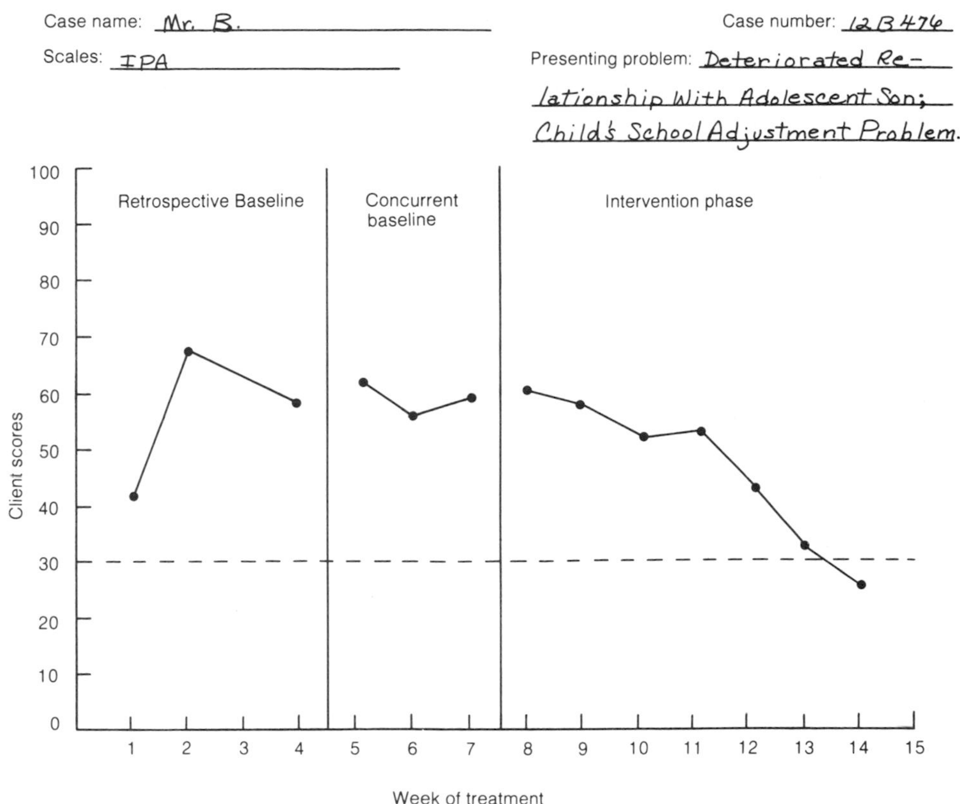

ethics. Some feel that baselining is used to verify experimental effects and that its use requires the act of depriving clients of the benefits of treatment for the questionable purpose of satisfying the whim of some scientist. It is my conviction that baselining sometimes raises such an ethical question, and when that occurs, the clinician must deal responsibly with the problem. However, it is often the case that debates about the ethics of baselining are conducted outside of the context of dealing with actual client situations and problems, and in those instances the debates are often reduceable to ideological disputes that take place between two or more opinionated and highly emotional individuals who are not at the moment concerned about the proper treatment of any specific client.

In actual practice, it is often the case that the therapist must meet with the client two or three or even more times before a formal intervention plan is developed or initiated. During such diagnostic, assessment, or planning stages, it is entirely possible to use one or more of the CMP scales as a means of obtaining concurrent baseline measures of the client's problem. In those cases where such procedures are feasible, failure to collect appropriate baseline measures will deprive the therapist of potentially important current information about the severity of the client's problem. Such failures also may severely limit the therapist's ability to clearly define and describe the magnitude of benefits or gains that arise as a consequence of the intervention that was used to help the client.

None of this is to say that concurrent baselining is essential in every case. For some clients, it is very important to begin some form of therapy or intervention immediately, and in those cases, the requirements of practice literally preclude the use of concurrent baselining procedures. While such practice realities must be recognized, it has been my experience that too many clinicians are too quick to claim that intervention must begin immediately for each and every client, and that baselining is therefore obviated for the overwhelming majority of all cases. While many cases do require prompt and immediate interventive action, the fact is that many do not. Also, when pressed for details, many therapists find it difficult to provide precise descriptions of treatment techniques that were used during the first two to four contacts with the client. In many cases, therapists do not dare initiate formal intervention simply because they have not yet had an opportunity to develop such on the basis of adequate and carefully considered information. The essential point is to be objective enough to recognize that concurrent baselining is feasible to some degree in a fairly large number of cases and to make good use of it as an aid to helping the client.

INTAKE AND SERVICE UNIT COLLABORATION

It is a practical reality in many agencies that the intake and service functions are made so distinct in terms of administrative organization that therapists from these two types of units sometimes do not have the opportunities to collaborate with one another. In such instances, the service unit therapist may have to exert special efforts to improve the chance of getting certain kinds of assessment data at the point of intake. In such cases, it may be necessary for the service unit therapist or the service unit supervisor to meet with members of the intake unit to determine whether some of the methods and procedures described here can be initiated within the intake unit. Barriers to effective professional collaboration may operate in the other direction as well.

In some agencies, it may happen that the intake counselors are initiating some of the methods described here because they want to provide sound and useful information to the service unit therapists.

In such cases, it may be necessary for the intake therapist or the supervisor of the intake unit to initiate some form of collaboration with service units to inform the latter of assessment and evaluation opportunities that were created within the intake unit. Because the CMP scales and the procedures described here are relatively new, special collaborative efforts may be needed in order for various types of social service units within the agency or clinic to make best use of these approaches to case assessment and evaluation.

ASSESSING CHANGE

Mrs. E. is a young mother who made an attempt at suicide. If the police had not found her, she would have died. When she was revived in the emergency room, she awoke crying and cursing the police and the doctors for saving her life. When she was released from the hospital, she entered therapy to alleviate a severe depression that was connected with a very destructive marriage. Figure 5–9

FIGURE 5–9

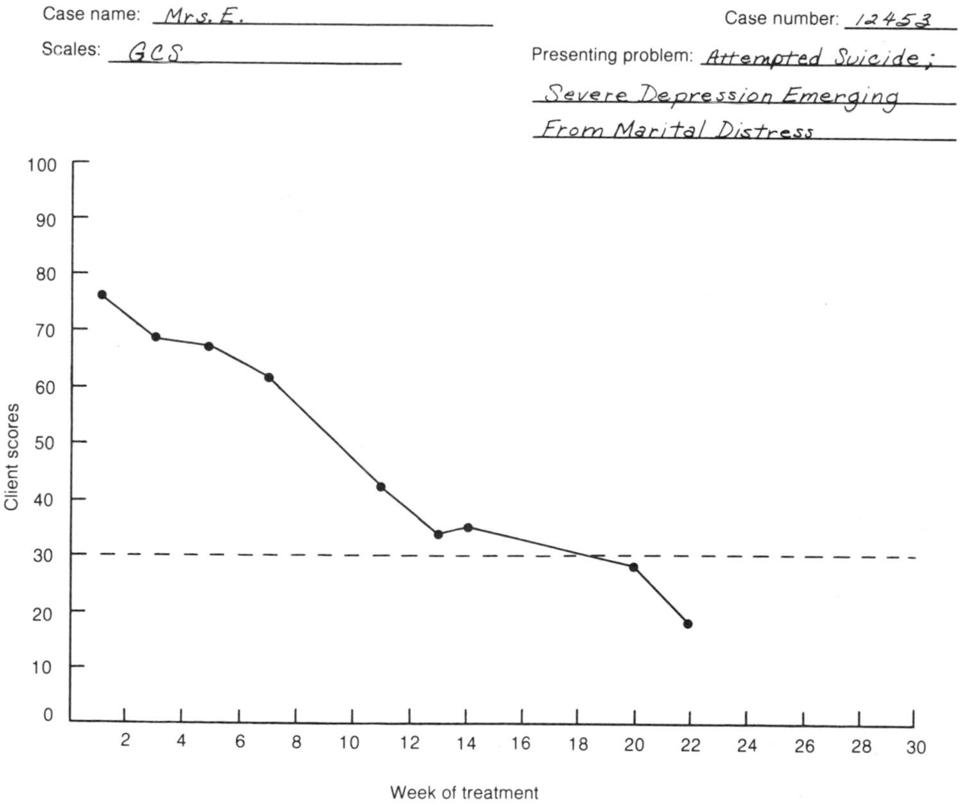

presents Mrs. E.'s depression scores, as measured by use of the GCS scale, from the onset of treatment until the case was closed.

Examination of Figure 5-9 strongly suggests that Mrs. E. made steady improvements, and when the case was closed, Mrs. E. no longer suffered from a clinically significant depression. There are no baseline measurements of Mrs. E.'s depression, but the overwhelming picture is one of dramatic positive change and eventual recovery. Mrs. E. gave me an anonymous copy of her autobiography, delivered by the therapist, and it provides convincing support for the picture that is represented by Figure 5-9. The final notes in this case indicate that Mrs. E. was successfully working on an advanced university degree and that she and her young son were having a good life together. The deceitful and destructive husband had been removed from Mrs. E. and her son by divorce.

On the basis of the evidence presented in Figure 5-9 and Mrs. E.'s autobiography, I am convinced that Mrs. E. did make a genuine and dramatic recovery and that her success was due in large part to the skill of the therapist. Yet there are some distinct problems with this case. If one wants to demonstrate that it was the therapy that was in part responsible for Mrs. E.'s recovery, these data are hopelessly inadequate for such a purpose. One could easily argue that Mrs. E. would have recovered on her own. Some form of baseline measure is absolutely essential to contradict such an argument. If baseline measures were available to show that Mrs. E.'s depression was stable at a very high level, the rapid drop in her depression scores that began with the onset of therapy would then suggest that it was treatment that was partly responsible for her recovery.

In spite of the logical difficulty involved in ascribing this success to the effect of treatment, the data shown in Figure 5-9 provide convincing evidence of marked positive change in the client's problem. As indicated earlier, that is the single most important focus of treatment. In a case such as this one, the therapist might have collected data for a retrospective baseline, and that would have been useful in helping to support the claim that Mrs. E.'s recovery was in part a *benefit* of practice.

Even though a baseline is not available for this case, the therapist's efforts to monitor Mrs. E.'s progress were extremely important. If at any time the GCS scores had shown a sudden turn for the worse, that would have provided the therapist with a signal that some event or circumstance within the treatment or elsewhere was creating a disturbance in the progress that was apparently being made. Such changes in score patterns provide essential clues and information that can help clinicians to recover or obtain therapeutic control over the problem that is being treated.

BASIC DESIGNS

In order to discuss and describe various treatment plans and strategies, it is often useful to employ some of the nomenclature found in the research literature. In the field of single-subject research, it is common to designate the baseline period by the letter *A* and to designate one or more intervention periods by a different letter. Thus, an *A*–*B* design would consist of a treatment plan that has one baseline period and one intervention period. The data shown in Figure 5–7 for Mrs. J. could be seen as an *A*–*B* design, those shown in Figure 5–9 for Mrs. E. represent a simple *B* design (no baseline), and those in Figure 5–8 for Mr. B. could be represented as an A_r–A_c–*B* design. In this third design the subscripts, r and c, are used to distinguish the retrospective baseline from the concurrent baseline. If a baseline subscript is omitted, it is assumed to be a concurrent baseline. By using this convention, the data for Mrs. J. in Figure 5–7 should be represented as an A_r–*B* design.

As seen in the case of Mrs. E., the simple *B* design is used only to monitor client change during the period of intervention. For the scientist who is seeking to study the relationship between client problems and methods of intervention, this design is worthless. However, the simple B design serves as a very powerful aid to professional practice for the clinician who wishes to monitor and assess client change and who wants to use such devices as the CMP scales to obtain feedback information about client problems and changes in those problems.

There are many instances in which the clinician would like to know whether some very specific intervention strategy (or perhaps a group of techniques) will in fact produce change in the client's problem. In such instances, the clinician must use, at a minimum, the *A*–*B* design, the A_r–*B* design, or the A_r–A_c–*B* design. These are very powerful designs for use in clinical practice, although scientists, for their purposes, would regard them as being rather weak. For clinical practice, these designs are extremely useful because they enable the therapist to use feedback to monitor client change, but they also provide a basis for determining whether observed change in the client's problem during the period of treatment was in fact *due to* the treatment strategy that was implemented. If kept in mind that the process of clinical practice literally *demands* some form of baselining by its very nature, it then becomes difficult to resist the conclusion that the simple *A*–*B* design should be regarded as the fundamental model of all forms of clinical practice.

Before leaving this section, caution is in order with respect to making use of the simpler *B* design. For certain types of cases, it must be regarded as the basic strategy that one must follow because the nature of the client's problem demands immediate action by the therapist. However, therapists should not too hastily decide that most of their treatment plans follow the *B* design. As pointed out earlier, there are many cases in which the initial contacts with the client consist of, or are devoted to, diagnosis, information gathering, and clinical analysis. Such periods might better be regarded as a form of baselining so that what might first appear to be a type of B design turns out, upon close inspection, to be an *A*–*B* design.

There are many other designs that can be used in clinical practice, but a discussion of them would go beyond the purpose of this manual. The basic purpose of this chapter has been to provide a brief introduction to some of the basic methods that can be used to monitor and evaluate the degree or magnitude of client problems through repeated administrations of one or more of the CMP scales. These basic monitoring techniques are sufficient to produce rather remarkable and beneficial changes in the conduct of clinical practice. After these basic methods have been used and mastered, the therapist may wish to consider other design options that are available. There are several excellent texts the reader may wish to consult for more extensive discussions of these topics (Hersen and Barlow, 1976; Fischer and Bloom, 1982; Rosenblatt and Waldfogel, 1982; Gottman and Leiblum, 1974; Sidman, 1960; Gambrill, 1977; and Kratochwill, 1978).

6 Reliability estimates for the CMP scales

This chapter presents detailed information concerning the reliability of the CMP scales. It begins with a brief discussion of the concept of reliability for the benefit of those who are not familiar with basic measurement theory; it explains the more common types of reliability that are used in measurement validation research; and it introduces the concept of measurement error. After presenting these background discussions, the remainder of the chapter provides the currently available reliability findings for each of the CMP scales.

THE MEANING OF RELIABILITY

As indicated in Chapter 1, if a measurement device is to have any clinical or scientific utility, it must be reliable. There are several ways to define the concept of reliability and there are several ways to compute an index of the reliability of a measurement tool. Conceptually, the most important definition of reliability is based on two theoretical concepts: true scores and error scores. Both of these concepts are theoretical in the sense that one can never obtain the absolute "true" score or the absolute "error" score. In spite of this limitation, these concepts provide a powerful basis for developing an extensive, highly effective theory of measurement error. An excellent discussion of this theory can be obtained from Nunnally (1978).

Suppose, for example, that the absolute true score for some specific client with respect to the ISE scale was a score of 38. Suppose also that this client completed the ISE scale on one occasion and obtained a score of 39. If 38 is the true score and 39 is the *observed* score, the difference between the two scores would represent the amount of *measurement error* involved in the use of the ISE scale on this one occasion for this particular client. Now, the true score of 38 will be denoted as T, the observed score of 39 will be denoted as O, and the error score of $39 - 38 = 1$ is denoted as E. These basic ideas can now be presented in the form of a simple but very important equation,

$$O = T + E, \tag{6.1}$$

which states that any observed score is equal to the true score plus the error score.

These concepts provide a basis for defining the concept of reliability. One very important definition states that reliability is based on the amount of error (actually, the lack of it) that is contained in an observed score for an individual or the amount of error that is contained in the observed scores for a population of individuals. If the amount of error is quite small, as judged by some well-defined criterion, then the measurement tool is said to be highly reliable. If, however, the amount of error is quite large, the measurement tool is then said to be unreliable.

In order to obtain any precision in describing the reliability of a measurement tool, it is necessary to quantify the above concepts. One way to do that is to find some way to compute the variance of the error scores, the true scores, and the observed scores—or to obtain reasonable estimates of those variances. If that can be done, and if the concept of variance is denoted by the symbol, s^2, the concept of reliability, denoted as r_{tt}, can be precisely defined as

$$\begin{aligned} r_{tt} &= s_t^2/s_o^2 \\ &= s_t^2/(s_t^2 + s_e^2) \end{aligned} \tag{6.2}$$

where s_t^2 is the true-score variance, s_e^2 is the error-score variance, and s_o^2 is the observed-score variance. This equation represents a formal definition of reliability which states that the reliability of a measurement tool is the proportion of the true-score variance that is contained within the observed-score variance. This proportion can range from 0.0 to 1.0, so that means that the range of any reliability estimate can be only from 0.0 to 1.0. If the proportion of true-score variance is quite large, the measurement tool is regarded as being highly reliable. If that proportion is quite small, the measurement tool is regarded as being unreliable.

RELIABILITY STANDARDS

Since reliability estimates can range from 0.0 to 1.0, a natural question arises as to how high must the reliability estimate be in order for the measurement tool to be regarded as having acceptable or good reliability. Obviously, a scale that has a reliability coefficient of .50 is better than one that has a coefficient of .40, but neither of these is regarded as being acceptable in applied work. Because of some misconceptions concerning the need for rigor in science, many might suspect that a measurement tool must have a higher reliability for scientific applications as compared to one that is to be used for clinical applications. (Interestingly, the reliability standards for scientific and clinical applications are just the opposite.)

It is usually accepted that a measurement tool that will be used for scientific work must have a reliability of .60 or greater, whereas one that is to be used in clinical practice must have a reliability of .80 or greater. The reasons for these differing standards arises from the fact that a great deal of scientific work capitalizes on the ability to average out errors of measurement by using large samples. Hence, useful results can be obtained by using measurement tools that have somewhat lower reliability. In clinical practice, one often uses a measurement tool to make decisions about, or in behalf of, a single individual, and in such applications it is apparent that sample data cannot be used to average out any measurement error. When making decisions about an individual, one must insist on having less tolerance for measurement error and the standards of reliability for such applications must therefore be more stringent.

THE STANDARD ERROR OF MEASUREMENT

The foregoing represents a brief introduction to the concepts of reliability and measurement error, and equation 6.2 provides a general definition of the coefficient of reliability. Reliability coefficients of one sort or another are the most common devices that are used to represent the reliability of a measurement tool or to describe its measurement error characteristics. However, most of the reliability coefficients that are used in applied work are based on the use of correlation coefficients in one way or another. Correlation coefficients are indexes that range in absolute value from 0.0 to 1.0, and they provide a very effective basis for characterizing the reliability of a measurement tool. However, one of the shortcomings of reliability coefficients that are based on correlations consists of the fact that they can be influenced by differences in the variance and standard deviation of a measurement scale that may occur from one sample or population to the next.

Another way to characterize the reliability of a measurement tool is to use a device called the standard error of measurement or SEM. A great advantage of the SEM is that its value is not influenced by differences in the variance and standard deviation of a measurement tool from one sample or population to the next (Helmstadter, 1964; Nunnally, 1978). Basically, the SEM is nothing more than an estimate of the standard deviation of the errors of measurement, E, shown in equation 6.1. However, instead of computing the SEM directly from the values of E that might be obtained from some specific sample, it is computed more efficiently as

$$SEM = s_o \sqrt{(1 - r_{tt})} \qquad (6.3)$$

where s_o is the standard deviation of the observed scores and r_{tt} is an estimate of the reliability of the measurement tool.

If a measurement tool has a very small SEM, it is judged to be a sound measurement device in terms of its measurement error characteristics. If it has a large SEM, it is judged as a poor measurement device in terms of its measurement error characteristics. The major disadvantage of the SEM is that there are no clear criteria for judging what is a small or large SEM. Generally speaking, a good measurement device, from a measurement error point of view, is one that has a large coefficient of reliability and a small SEM in relation to the overall range of possible scores. In spite of this interpretational weakness of the SEM, it is an important way of representing one aspect of the reliability of a measurement tool. For that reason, the SEM, as well as the reliability, is computed and reported for each of the CMP scales.

TYPES OF RELIABILITY

In applied work, there are basically three different types of reliability measures that are most commonly employed in standardizing a new measurement device. These are referred to as internal consistency measures, test-retest measures, and parallel form (or alternate form) measures. A detailed technical discussion of these three types of reliability is not needed for the purposes of this manual, but some comment and clarification is in order.

Split-half reliability

Split-half reliability is probably the most common or widely known form of reliability estimate. It is computed by using half the items on a scale to compute a score and then using the remaining items to compute a second score. These two scores are then correlated for a large sample of individuals, and the result is referred to as the split-half correlation. This split-half correlation is then corrected for test length by use of the Spearman-Brown Prophecy formula (Nunnally, 1978), and the corrected value is then referred to as the split-half estimate of reliability.

Actually, the split-half method of estimating the reliability of a scale turns out to be only an *estimate* of another internal consistency measure, called coefficient alpha, that is much more powerful than the split-half method. In the early phases of the validation research for the CMP scales, I used the split-half method of estimating the reliability of the scales because I did not know that a better method was available. Once it was learned that alpha was a better device, the split-half method was abandoned entirely, and all subsequent estimates of reliability for the CMP scales were based on alpha.

The alpha coefficient

Coefficient alpha was chosen as the primary means of estimating the reliability for each of the CMP scales because it has a number of highly desirable characteristics. It is an internal consistency measure of reliability that is based on all the interitem correlations for a particular scale, it is the mean of all possible split-half reliabilities, and it provides a direct estimate of the alternate form reliability that would be obtained if an equally good alternate form of a particular scale were available. Another desirable feature of alpha arises from the fact that an alpha coefficient of .90 or greater provides direct evidence to support the claim that a particular scale is a unidimensional measurement tool (Nunnally, 1978).

Coefficient alpha is also very easy to compute. If the correlation between any pair of items on a scale is denoted as r and the arithmetic mean of all possible pairwise item correlations is denoted as $\bar{r}$ (the number of such correlations is $N = (k^2 - k)/2$ where k is the number of items on the scale), then coefficient alpha can be computed as

$$Alpha = k\bar{r}/(1 + (k-1)\bar{r}) \qquad (6.4)$$

It should be noted that equation 6.4 is technically known as the "generalized Spearman-Brown" or GSB formula, and that alpha is really a specific derivation from the GSB formula that was developed by Cronbach (1951). Because of the remarkable similarity between Cronbach's alpha and the GSB formula, I have taken the liberty of referring to equation 6.4 as the alpha coefficient. The major point is that equation 6.4 was used to compute the reliability for each of the

CMP scales, and that is the specific statistic that is referred to as alpha in the technical papers, cited later in this chapter, that report the author's and his colleagues' efforts to validate the scales.

Test-retest reliability

The second major type of reliability estimate used in applied work is referred to as test-retest reliability. Test-retest reliability is computed as the correlation between a set of scores obtained from a single scale on two separate occasions. It presumably represents the stability of a measurement device over time, but it requires the critical assumption of no change in the measured characteristic (except for variations due only to measurement error) from one occasion to the next. If the test interval is quite long, the likelihood is increased for obtaining real change in the measured characteristic, and the effect of such change is one of producing an underestimate of the true reliability of the scale.

One way to possibly compensate for this problem is to make the test interval short enough to ensure that real change in the measured characteristic does not occur. Unfortunately, if the test interval is too short there arises a risk or likelihood that scores obtained from the second administration of the scale will be influenced by the respondent's recollection of answers provided on the first occasion. Such memory effects usually produce an exaggerated estimate of the true reliability of the measurement device that is being evaluated.

In applied work, it is often very difficult to determine what time interval between two test occasions will simultaneously minimize the risk of underestimating the reliability of a scale because of the effects of real change in the measured characteristic and minimize the risk of an exaggerated reliability estimate that arises because of memory effects. Because of these problems it is my conviction that estimates of test-retest reliability are, at best, of dubious value and that this type of reliability has been given entirely too much emphasis. At any rate, very little effort has been made to compute estimates of test-retest reliability for the CMP scales. Such estimates have been produced in a few instances in response to peer reviewers of journal articles.

Alternate form reliability

The third major type of reliability estimate is referred to as alternate- or parallel-form reliability. If one really needs two or more parallel or alternate forms of a particular scale, an estimate of this type of reliability is essential. However, alternate forms of the CMP scales are not currently available and it is thus not possible to compute alternate form reliabilities for these scales. The alpha coefficients reported later in this chapter are estimates of what the alternate form reliability would be for the CMP scales if equally good alternate forms of the scales were available.

PARALLEL FORMS OF THE CMP SCALES

There may be some occasions in which a clinician or researcher would like to have two different forms of one or more of the CMP scales. On such occasions it may be tempting to create two forms of a scale by randomly assigning half of the items to one form and the other half of the items to a second form. This practice is not recommended for several reasons.

Suppose that the value of alpha for some specific scale is .90 and it has 25 items. If equation 6.4 is solved for $\bar{r}$ by using these values, it will be seen that $\bar{r} = .26471$, approximately. If it is assumed this will be the average correlation among the first 12 items as well as the last 13 items, one can then estimate the reliability of these two half-length scales. By putting $\bar{r}$ back into equation 6.4 it will be seen that the alpha coefficient for the 12-item scale will be .8120, and for the 13-item scale it will be .8239. Now, under the assumption that the two forms of the scale were truly parallel and that they would correlate perfectly with one another in the absence of measurement error, the maximum possible *observed* correlation between these two half-length scales would be equal to the square root of the product of their reliabilities, that is, .8179. It is my judgment that this is not good enough. For applied clinical work, this value should be .90 or better.

For research applications based on large samples, it may be acceptable to split the scales into alternate forms, but even here there are serious risks of underestimating the sizes of the correlations that one is trying to investigate and describe. If alternate forms of the CMP scales are to be created by dividing the scales into half-length forms, it is strongly recommended that the shortened forms be separately validated. If alternate forms of a measurement device really are essential for some specific clinical or research application, it is also essential that the forms be truly parallel. Support for such a claim is difficult to establish if the alternate forms correlate with one another at less than .90, and it is very doubtful that half-length forms of the CMP scales will measure up to such standards.

A second reason for advising against splitting the CMP scales into alternate forms arises from the fact that the upper limit of the validity of any scale is established as the square root of its reliability. Very few scales ever have validity coefficients that approach this upper limit, but the importance of this relationship is clear. The validity of any scale is determined, in part, or limited by its reliability. If the validity of a scale is .75 with a reliability of .90, a reduction in reliability to .80 will have the effect of reducing the validity to .60. In short, it is important to protect the existing validity of the CMP scales (described in the next chapter) by protecting their reliabilities. The latter (hence, the former) is best achieved by always using the full-length scales as shown in Chapter 1.

THE CMP RELIABILITY FINDINGS

The purpose of the above materials has been to provide background information concerning the concepts of reliability and measurement error and to indicate how the reliability estimates for the CMP scales were generally obtained. In the remainder of this chapter, detailed findings are presented to describe and to report current reliability estimates for each scale.

There are currently 16 research projects or study samples that relate to one or more aspects of the validation of the CMP scales, and two others are in progress. Each of these projects is numbered below, and citations are given for the research reports that were prepared from the separate projects. A brief description of each study is provided with the citations. Detailed information concerning the reliability and validity of the CMP scales can be obtained from the cited reports, but the findings concerning the reliability of the scales are summarized in two separate tables later in this section. The research projects and reports are listed, cited, and described as follows.

1. Hudson and Glisson (1976). An initial study to partially validate the IMS scale.
2. Hudson and Proctor (1976a, 1977). An initial study to partially validate the GCS scale.
3. Hudson and Proctor (1976b). A study to partially validate the ISE scale.
4. Giuli and Hudson (1977). A study to partially validate the CAM and CAF scales.
5. Bartosh (1977). A study of depression and family stress among graduate students.
6. Hudson, Hess, and Matayoshi (1979). A study of personal and social functioning and their relationship to academic performance in a private boarding school in Honolulu.
7. McIntosh (1979). A study designed to partially revalidate the ISE scale and to obtain an improved estimate of discriminant validity.

8. Hontanosas, Cruz, Kaneshiro, and Sanchez (1979). A study conducted with me to examine the incidence and severity of spouse abuse among females on a state university campus.
9. Murphy (1978); Murphy, Hudson, and Cheung (1980); Hudson and Murphy (1980). A study of personal and social functioning and their relationship to ethnicity and stages of the family lifecycle.
10. Byerly (1979). A study that compared inpatients, outpatients, and normals with respect to three self-rating depression inventories.
11. Hudson, Wung, and Borges (1980). A study to partially validate the IPA scale.
12. Hudson, Acklin, and Bartosh (1980). A study to investigate the reliability and validity of the IFR scale.
13. Nurius (1982); Hudson and Nurius (1981). A study of personal and social functioning and their relationship to several types of sexual activity and preferences.
14. Hudson, Hamada, Keech, and Harlan (1980). A study of the reliability and validity of the GCS, Zung, and Beck depression scales.
15. Cheung and Hudson (1981). A study conducted to revalidate the IMS scale.
16. Hudson, Harrison, and Crosscup (1981). A study to investigate the reliability and validity of the ISS scale.
17. Hudson, Abell, and Jones (1982). This study is currently in progress and is being conducted to obtain improved estimates of the validity of the ISE scale.
18. Nurius and Hudson (1982). This study is currently in progress and is being conducted to investigate the reliability and validity of the IPR scale.

The research findings obtained from the above study samples were summarized to obtain a single best estimate of the reliability for each of the CMP scales. When two or more estimates of alpha were available for one of the scales, an overall best estimate was computed as the mean of the separate estimates. An overall best estimate of the SEM was obtained by computing the mean of the SEM estimates obtained from the separate study samples. The reliability data from the separate studies are presented by project number in Table 6–1 for the GCS, ISE, IMS, ISS, IFR, CAM, and CAF scales because they are the ones for which multiple estimates are available. The reliability and SEM for the IPA scale is based on only one sample, and those data, along with the best estimates for the other scales, are presented in Table 6–2.

TABLE 6-1
Multiple reliability estimates for seven of the CMP scales

Project number	Sample size	Alpha	SEM
GCS scale:			
2	132	.92	4.24
8	398	.90	3.96
9	378	.89	3.95
10	219	.89	5.23
12	120	.96	4.32
13	693	.89	3.82
14	200	.96	4.32
ISE scale:			
3	124	.92	3.51
7	59	.91	3.87
8	398	.94	3.66
9	378	.93	3.85
11	93	.95	3.73
13	693	.94	3.55
IMS scale:			
1	124	.96	3.04
8	398	.96	3.89
9	378	.95	3.67
13	693	.95	3.73
15	110	.97	4.59
16	100	.94	5.09
ISS scale:			
8	398	.92	4.29
9	378	.92	4.08
11	59	.94	4.70
13	693	.91	4.13
15	110	.93	4.98
16	100	.92	3.27
CAM scale:			
4	664	.94	5.06
6	408	.93	4.08
CAF scale:			
4	664	.95	5.16
6	408	.95	3.96
IFR scale:			
5	198	.91	4.58
12	120	.98	2.40
14	200	.97	3.96

The data reported in Tables 6-1 and 6-2 provide strong and convincing evidence to support the claim that each of the CMP scales is a highly reliable measurement device and that each of the scales has a very small SEM in relation to scale scores that can range from 0 to 100. Although reliability data are not yet available for the IPR scale, it is strongly suspected that it will be found to be a highly reliable

TABLE 6-2
Best reliability estimates for the CMP scales

Scale name	Sample size	Alpha	SEM
GCS	2140	.92	4.26
ISE	1745	.93	3.70
IMS	1803	.96	4.00
ISS	1738	.92	4.24
IPA*	93	.97	3.64
CAM	1072	.94	4.57
CAF	1072	.95	4.56
IFR	518	.95	3.65
IPR	n.a.	n.a.	n.a.

n.a. = not available.
* See Hudson, Wung and Borges (1980).

measurement device. While there is good reason to be confident of such an outcome, those who choose to use the IPR should proceed with caution until reliability findings are made available. Those data will be forthcoming in the near future.

7 Evaluating the CMP validities

This is the most important chapter of the manual. It presents the currently available evidence to support the claim that the CMP scales really do measure what they are supposed to measure. If the scales do not measure what they are supposed to measure, they are not valid, and they cannot have any clinical or scientific utility. Before presenting the findings concerning the validity of the CMP scales, background information is provided to describe the methods that were used to investigate the CMP validities.

THE MEANING OF VALIDITY

The fundamental definition or meaning of validity was implied in the above paragraph. The validity of any measurement tool consists of its ability to measure what it was designed to measure. An older view of validity, one that is no longer followed, was that a measurement tool was either valid or not valid. There were no shades of grey. Such a view has been disspelled because it is now recognized that one measurement device may do a much better job than another of measuring the same construct and both may be judged as being valid. Validity, like reliability, is a matter of degree, so it is important to recognize that any measurement tool must be judged as being more or less valid in relation to one or more well defined criteria. The validity of a scale is often, but not always, described by computing some form of a *validity coefficient.* Such coefficients are nearly always derived as some form of percentage estimate or as some form of correlation coefficient, so validity coefficients nearly always have a theoretical range of values from 0.0 to 1.0.

Another way to define validity is to say that a measurement tool is valid if it will do an acceptable job of fulfilling some function or task. Thus, a scale can be valid for one purpose, task, or function but not others. If a personnel classification test does a good job of predicting who will be a good clerk-typist but a poor job of predicting success as a space project engineer, it is seen as being valid for the first task but not for the second. In these terms, it should be stated from the outset that the CMP scales are not valid devices for classifying clients or other respondents into various nosological categories. They simply were not designed to do that. They also do a very poor job of predicting athletic ability, the weather, and changes in the stock market! The fundamental question in this chapter is whether they do a good job of measuring the degree or magnitude of the problems they were designed to measure.

VALIDITY STANDARDS

As was noted briefly in the previous chapter, validity and reliability are related to one another. Specifically, the absolute upper limit of the validity of a measurement tool is established as the square root of its reliability. If some validity coefficient is denoted as V, then the upper limit of that value will be $V = \sqrt{r_{tt}}$ and any computed validity coefficient will have the value of $V \leqq \sqrt{r_{tt}}$. For example, if the reliability of some measurement tool is $r_{tt} = .64$, then the value of any computed validity coefficient will be $V \leqq \sqrt{r_{tt}} \leqq \sqrt{.64} \leqq .80$.

Although the square root of the reliability sets the upper limit of validity for any particular scale, in practical applications it is nearly always found that validity coefficients are much smaller than their theoretical upper limits; often less than two thirds or even half of that value. What, then, should be considered as a standard for defining good validity? It is not easy to pin down a single figure as a criterion, but it is useful to know something about typical ranges for validity coefficients and to remember that they tend to be much smaller than reliability coefficients.

According to Downie and Heath (1967), validity coefficients tend to range from a low of about .40 to a high of about .60 with a median of about .50. This represents a typical or common range and these figures should not be regarded as lower or upper limits. If this median figure is correct, it indicates that any scale with a validity coefficient greater than .50 is among the best 50 percent of all scales in terms of its ability to measure what it is supposed to measure. Thus, the median validity provides one basis for judging the quality of a measurement tool in relation to others.

Although the median validity coefficient is very useful, for purposes of this manual and for purposes of establishing some criterion for judging the CMP scales, it was decided to regard any measurement tool that has a validity coefficient in excess of .60 as being an unusually good one. In short, the upper end of the range, reported by Downie and Heath (1967), was adopted as a standard of "good" validity since few measurement tools appear to have validities that exceed that value.

TYPES OF VALIDITY

There are many different types of validity, and several of them are described in the following sections of this chapter. In terms of the purposes for which the CMP scales were designed, the most important types of validity are referred to as content, criterion, construct, and factorial validity. Any standardized measurement tool that is to be used for both clinical and scientific applications in the helping professions should be valid in terms of each of these types of validity. The following sections present a discussion of the above types of validity and then describe how each was investigated for the CMP scales. That material is followed by a description of the currently available evidence concerning the validities of the scales.

FACE VALIDITY

Face validity is without doubt the weakest form of validity. It consists largely of the unsupported claim that a measurement tool, "on the face of it," does what it is supposed to do. Face validity is so weak that it may be wise to adopt a policy of granting *any* measurement device an acknowledgement of face validity and thereby dispensing with the entire matter. By doing so, nothing is gained or lost and one can then get on with the task of looking at the forms of validity that matter. Thus, it is of little consequence to say that all of the CMP scales have good face validity.

CONTENT VALIDITY

Content validity, in a very great way, is central to the issue of validity from a scientific and a clinical point of view. Given that one wishes to measure some specific construct, attribute, or personal (or interpersonal) relationship problem, it may be presumed that there is a potentially infinite number of items that could represent some part or aspect of the disorder to be measured. If such a population of items could be defined, it would be highly desirable to develop a specific measure of the disorder by taking a random sample of items from the population. The laws of chance would reasonably ensure that a random sample of such items would do a good job of representing the content of the population of items.

In practice, it is literally impossible to create or define the total population of items that could represent different measures of a single construct, attribute, or problem. It is therefore impossible to draw a random sample of items from that population to ensure that the item sample comprising the scale actually represents the item population. Thus, content validity is an idea that one uses and an ideal that one pursues in confronting the task of developing a content-valid scale.

How, then, does one judge or evaluate the content validity of the CMP scales? Because of the limitations that are inherent in any effort to define a universe or population of items comprising the content domain for any construct, an assessment of content validity ultimately reduces to an act of judgment on the part of the clinician and scientist about each of the items in a scale. To the extent that a specific measurement tool is judged to have items that adequately represent the content domain, it is thereby judged as having good content validity.

One way to approach an assessment of the content validity of a scale is to measure the degree of consensus among competent judges as to whether each item on the scale is a member of the population of items that comprises the content domain. If high consensus is obtained, one could then claim that the scale has high content validity. If such consensus cannot be obtained, then it is doubtful that a claim for good content validity could be defended. At this time, no formal data are available to represent the degree of consensus among judges as to the content validity of the CMP scales. Such data are badly needed, and a formal study to obtain such data is planned.

A second way to partially examine the content validity of a scale is to examine its factorial validity. Factorial validity is investigated through the use of a special form of an analytic technique known as "factor analysis" that is reduceable to a form of item analysis. The factorial validity of the CMP scales is described in a separate section, and a final judgment concerning the content validity of the CMP scales will be made after presenting and discussing the factorial validity findings.

An important test for individual clinicians to make for themselves is to read carefully each of the items on each of the CMP scales. If an item is judged as definitely measuring the problem defined for that scale, place a plus sign (+) beside the item. If an item is judged as definitely measuring some other construct than the problem defined for the scale, place a minus sign (−) beside the item. The number of plus signs divided by 25 will give a crude index of content validity in terms of the individual clinician's own judgment for each scale.

CRITERION VALIDITY

Criterion validity refers to the ability of a measurement tool to explain or account for some well-specified criterion. Criterion validity is usually assessed in terms of some type of correlation measure. If the observations on the measure to be validated are taken at some point in time, T_1, and the observations with respect to the criterion measure are taken at a later point in time, T_2, the correlation between the two measures is referred to as an index of *predictive* criterion validity. If the observations are taken at the same point in time with respect to the criterion measure and the measure that is to be validated, the correlation between the two measures is then referred to as an index of *concurrent* criterion validity. There are no data currently available to assess the predictive criterion validity of the CMP scales.

There are two different types of concurrent criterion validity. The first type will be referred to as concurrent *instrument* validity. Concurrent instrument validity is examined by correlating a new measurement tool with one that presumably has already been shown to be a valid measure of some construct. Very little emphasis has been given to this form of validity, and thus far it has been estimated only for the GCS scale. In a recent study by Hudson, Hamada, Keech, and Harlan (1980), the GCS scale was correlated with the Beck (Beck, Ward, Mendelson, Mock, and Erbaugh,1961) and Zung (1965) depression scales. When the Beck scale was treated as the criterion, the GCS concurrent instrument validity was found to be .85 for a sample of 200 persons and .76 for another sample of 120 persons. When the Zung scale was used as a criterion, the GCS concurrent instrument validity was found to be .92 and .81 for the two samples, respectively. In a study conducted at Western Michigan University, Byerly (1979) reports concurrent instrument validities for the GCS that range from .58 to .80 when the Beck scale is used as a criterion measure, and the range was from .56 to .78 when using the *Depression Adjective Checklist* or DACL (Lubin, 1967) as a criterion. The wide range of validity coefficients in the Byerly study probably arises from an attenuation of the correlations that arose because of a truncation of range.

The second type of concurrent criterion validity is based on a procedure that is often referred to as the *known groups* method. In using the known groups method of assessing concurrent validity, one tries to obtain at least two groups of people wherein one is known or judged as having more of the attribute, characteristic, or problem than the other. Individuals in each group are then asked to complete the measurement device that is to be evaluated. If the measurement tool is a valid measure of the construct or attribute, the difference between the two known groups with respect to their mean scores should be statistically significant. Moreover, if the two groups are treated as the independent variable and the scale that is to be validated is treated as the dependent variable, the point-biserial correlation between group status and the scale can be regarded as a coefficient of validity. In other words, the fundamental thesis behind this procedure means that if the measurement tool is valid, it should do a reasonably good job of discriminating between the criterion groups.

Because the fundamental evidence concerning the validity of a scale, insofar as this procedure is concerned, consists of the ability of the scale to discriminate between two or more known criterion groups, I have taken the liberty of referring to this evidence as discriminant validity. Thus, the point-biserial correlation between group membership and the scale that is being investigated will be referred to as a coefficient of discriminant validity.

DISCRIMINANT VALIDITY

Thus far, discriminant validity has been well established for each of the first eight CMP scales, and a validity study for the IPR scale is currently in progress (Nurius and Hudson, 1982). In all cases, the known criterion groups were created by asking experienced therapists (most of whom were social workers and psychologists) to divide their clients into three groups: (1) those who definitely had a clinically significant problem in the area being studied; (2) those who were definitely free of a clinically significant problem in the area being investigated; and (3) those about whom the therapists were uncertain. In short, the independent clinical diagnosis of experienced professional therapists from among many different settings (public and private social welfare agencies and private practice) were used as the basis for creating the known criterion groups against which the discriminant validity of the CMP scales would be evaluated. After dividing clients into the above three groups, clients in the two known groups were asked to complete a research questionnaire package that contained the particular CMP scale that was being evaluated.

After the clients completed the research questionnaires, the data were analyzed by conducting a one-way analysis of variance in which the scale being evaluated was treated as the dependent variable and membership in the criterion groups was treated as the independent variable. The primary findings concerning the discriminant validity of the CMP scales are shown in Table 7-1 where the point-biserial correlation between group membership and the scale is treated as a coefficient of discriminant validity. Also shown in the table is the project number associated with the list of research projects that was presented in Chapter 6, the F-ratio associated with the test of significance of each point-biserial correlation, and the probability estimate of the risk of Type I error. Further details concerning the conduct of these validity studies can be obtained from the research reports that were cited in Chapter 6.

Examination of the summary findings shown in Table 7-1 strongly suggests that the CMP scales have good to outstanding discriminant validities as judged against the validity standards discussed earlier in this chapter. While these findings indicate that the CMP scales are valid for use in clinical and research applications as discussed in previous chapters, additional validation work is currently in progress, and more will no doubt be carried out in the future. A study by Hudson, Abell, and Jones (1982) is being conducted to obtain improved validity data for the ISE scale, and Nurius

TABLE 7-1
Discriminant validity findings for the CMP scales

Scale name	Project number	Discriminant validity	F-ratio	*P* less than
GCS	14	.74	178.09	.0001
ISE	3	.52	—*	.001
IMS	15	.82	200.36	.0001
ISS	16	.76	136.18	.0001
IPA	11	.88	209.95	.0001
CAM	4	.86	103.92	.0001
CAF	4	.87	115.11	.0001
IFR	12	.92	458.58	.0001
IPR	n.a.	n.a.	n.a.	n.a.

n.a. = not available.

* The original study data were lost and it is not possible to report the F-ratio. Also, this study is based on self-report data entirely, and the criterion groups were not formed in the same manner as with the other studies. See the original report for details.

and Hudson (1982) are conducting a study to obtain initial validity data for the IPR scale. (Both of these research reports should be available within a year after the publication of this manual.)

Thus far it has been argued that if a scale is a valid measure of the personal or interpersonal relationship problem it was designed to measure, it should do a good job of discriminating among two or more groups that are known to have different amounts of the problem being examined. The data shown in Table 7-1 indicate that the CMP scales do a good job of discriminating between such criterion groups. However, it can also be argued that if a scale is a valid measure of some construct or problem, it should do a better job of discriminating between appropriate known criterion groups than one or more other scales that were designed to measure some other construct or problem. For example, if the ISS is a valid measure of sexual discord, it should do a better job of discriminating two sexual discord criterion groups than either the IMS scale or some other scale that was designed to measure, say, the degree of a liberal or conservative orientation toward human sexual expression, for example, the *Sexual Attitude Scale* or SAS (Hudson, Murphy, and Nurius, 1981).

In several studies that were carried out to obtain discriminant validity data, respondents completed one to three other scales besides the one that was being assessed. Thus, the data obtained from these other measures can be used to further investigate the discriminant validity of the CMP scales. That is, most of the CMP scales had to compete against one or more other scales in terms of the ability of each to discriminate between appropriate known criterion groups. This procedure was used for the GCS, IMS, ISS, IPA, and IFR scales, and the findings are summarized in Table 7-2. The procedure consisted of computing a discriminant validity coefficient for each of the scales that the group members completed. The point-biserial correlations and their squared values are shown in Table 7-2.

When examining the findings shown in Table 7-2, it should be remembered that a comparison of correlations can be misleading because they do not represent an equal-interval metric. However, their squared values represent an equal-interval metric, so when trying to judge the relative discriminant validity of two or more scales, it is important to compare the squared values of the point-biserial correlations.

In judging the comparative validities for the GCS, it should be noted that it is being compared to two other well-known depression scales and one nondepression scale. The squared validity coefficients show the GCS to be as good as the Beck and Zung scales and

TABLE 7-2
Comparative discriminant validities for the CMP scales

Criterion groups	Scale name	r_{pb}	r^2_{pb}	F-ratio	*P* less than
Depression					
problem groups	GCS	.74	.55	178.09	.0001
	Beck	.70	.49	121.89	.0001
	Zung	.72	.52	143.80	.0001
	IFR	.50	.25	44.88	.0001
Marital					
discord groups	IMS	.82	.68	200.36	.0001
	ISS	.66	.43	69.78	.0001
	SAS	.08	.01	0.56	.4546
Sexual					
discord groups	ISS	.76	.58	136.18	.0001
	IMS	.52	.27	36.66	.0001
	SAS	.16	.03	2.51	.1163
Parent-child					
discord groups	IPA	.88	.78	209.95	.0001
	ISE	.45	.20	15.40	.0002
	ISS	.46	.21	10.46	.0025
Family					
stress groups	IFR	.92	.84	458.58	.0001
	GCS	.69	.47	74.85	.0001
	Beck	.55	.31	36.34	.0001
	Zung	.69	.48	75.06	.0001

about twice as good as the IFR scale in terms of its ability to discriminate the known criterion groups.

The comparative validities for the IMS scale show it to be much better (.68 − .43 = .25) than the ISS and about 70 times better than the SAS scale (Hudson, Murphy, and Nurius, 1981) in terms of its ability to discriminate between the known marital discord criterion groups. The relative closeness of the IMS and ISS discriminant validities no doubt arises from the fact that persons who have serious marital problems are also very likely to have problems in the sexual component of their relationship. These data suggest that it is difficult to have a satisfactory sexual relationship if the nonsexual component of the relationship is in trouble. The SAS scale is not part of the CMP, and it was developed to measure the degree of a liberal or conservative orientation toward human sexual expression. These data suggest that marital distress has little to do with one's attitudes toward sexual expression.

The comparative validities for the ISS scale show it to be about twice as good as the IMS scale and about 20 times as good as the SAS scale in terms of its ability to discriminate between known sexual discord criterion groups. These data suggest that it is easier to have a good marriage in spite of the presence of sexual discord and that sexual attitudes have little to do with sexual discord.

The data shown in Table 7–2 indicate that the IPA scale is about three times as effective as the ISE and ISS scales in terms of its ability to discriminate known parent-child problem groups. Also, the IFR scale does a much better job of discriminating known family stress groups than does three different measures of depression.

The essential findings shown in Table 7–2 are that the scales do a good job of measuring what they are supposed to measure and that they do a much better job than several other scales that were allowed to compete with them. Although such data are not currently available for the ISE, CAM, CAF, and IPR scales, this evidence considerably strengthens the case for claiming that the CMP scales have good discriminant validity.

CONSTRUCT VALIDITY

Construct validity refers to the ability of a measurement tool to measure the specific construct it was designed to measure. It is very much akin to content validity. However, content validity refers largely to the sampling of the construct domain and the *construction* of a measurement tool while construct validity refers to the *performance* of the device with respect to theoretical expectations.

In a classic paper by Campbell and Fiske (1959), construct validity was defined as the ability of a measure to correlate well with other variables that are theoretically highly related to the construct, but it also refers to the ability of a measure to correlate poorly with yet other variables that are theoretically unrelated to the construct. The first type of performance was called *convergent* validity, and the second was referred to as *discriminant* validity. Thus, convergent and discriminant validity are two different types of construct validity. Note carefully that discriminant validity, as defined here, is quite different from the type of known-groups discriminant validity that was discussed in the previous section.

The hallmark of construct validity consists of making strong or weak theoretical predictions and then finding that observed correlations are consistent with the theory. This is not a theory-testing procedure. Rather, the theoretical predictions serve as the criteria against which one judges a measure as having (or not having) construct validity. The basis for examining construct validity is derived directly from its definition. According to Campbell and Fiske (1959), discriminant validity consists of the finding that a measure has a low correlation with other variables it should theoretically be unrelated to, and convergent validity consists of the finding that a measure has a high correlation with other variables it should be theoretically related to. The demonstration of both convergent and discriminant validity is considered as evidence in support of the construct validity of a measurement tool.

In order to examine the construct validity of the CMP scales, it was necessary to develop at least some general a priori hypotheses concerning their relationships with one another and with a number of other variables. To the extent that data were available that could be used to examine the construct validity of the CMP scales, an effort was made to specify three levels of prediction. The first prediction was that each of the CMP scales would correlate the lowest with a set of basic social background variables such as age, sex, ethnicity, marital status, family size, and so on. These low levels of predicted correlation were made because it is believed that the types of personal and interpersonal relationship problems that are measured by the CMP scales actually have little to do with "who we are" as represented by background social characteristics. Because members of all social status groups are vulnerable to the problems represented by the CMP scales, the correlations between the scales and a set of social status variables should be very small. The social status variables will be referred to as class I criterion variables when examining construct validity.

The second prediction was that for each of the CMP scales there are a number of variables that are expected to have only moderate correlations with the particular scale to be evaluated. This list of variables will vary somewhat from one scale to the next, and it is necessary to specify that list separately for each of the CMP scales. These variables were obtained in most of the CMP studies by using a short self-report questionnaire called the *Psycho-Social Screening Package* or PSSP scale. A copy of the PSSP is shown as Figure 7-1, and it is described more fully in appendix B. In addition to measures taken from the PSSP, some of the CMP scales will be used as midrange criterion variables. Variables that are expected to have only moderate correlations with the CMP scales will be referred to as class II criterion variables when examining the construct validity of the scales.

The third prediction was that for each of the CMP scales there would be a group of variables that would have the highest correlations with the scales. Some of these variables will be taken from the PSSP and some will consist of other CMP scales. This group of variables will be referred to as class III criterion variables when examining the construct validity of the CMP scales.

In order to be as explicit as possible about stating the a priori hypotheses that were developed to examine the construct validity of

FIGURE 7-1

PSYCHO-SOCIAL SCREENING PACKAGE (PSSP)

This questionnaire provides you a means of reporting problems that you may be having as an individual or in your relationships with your family and others whom you know or work with. It is not a test, so there are no right or wrong answers. Answer each item as carefully and as accurately as you can by placing a number beside each one as follows:

1 Rarely or none of the time
2 A little of the time
3 Some of the time
4 Good part of the time
5 Most or all of the time
6 This item does not apply to my situation

Please begin.

1. I feel depressed ______
2. I have a low sense of self-esteem ______
3. I feel unhappy ______
4. I feel afraid ______
5. I feel anxious ______
6. I feel nervous ______
7. I have disturbing thoughts or fantasies ______
8. I have problems with my own anger or rage ______
9. I have nightmares ______
10. I have a problem with my own sense of identity ______
11. I have problems with my personal sex life ______
12. The quality of my work is poor ______
13. I have problems in my relationships with friends ______
14. I have problems in my relationships with people I work with ______
15. There are problems in our family relationships ______
16. I have problems with one or more of my children ______
17. My partner and I have problems in our relationship ______
18. My partner and I have problems with our sexual relationship ______
19. My mother and I have problems in our relationship ______
20. My father and I have problems in our relationship ______

the CMP scales, the class I and class III criterion variables are listed in Table 7-3 for each of the CMP scales. All other variables taken from the PSSP and the CMP will be treated as class II criterion variables.

Although it was possible to construct the explicit hypotheses identified in Table 7-3, it must be recognized that they are not based

TABLE 7-3
Criterion variables for the CMP construct validities

Class I criterion variables for all CMP scales:
Age
Sex
Ethnic status
Education
Income
Marital status
Family size
Class III criterion variables for the GCS scale:
Zung depression scores
Beck depression scores
Clinical status
Depression—PSSP
Self-esteem—PSSP
Happiness—PSSP
Sense of identity—PSSP
ISE scores
Class III criterion variables for the ISE scale:
Clinical status
Depression—PSSP
Self-esteem—PSSP
Happiness—PSSP
Sense of identity—PSSP
GCS scores

on strong theory (Platt, 1964). Thus, it was fully expected that some of the specific predictions would fail. The best that could be expected was that the predictions identified in Table 7-3 would hold *in general* and that the average correlation between each scale and each of the three classes of criterion variables would be strong enough to show that the CMP scales were valid as a construct. Consequently, the construct validity findings were summarized by averaging all the correlations between a specific CMP scale and each of the class I criterion variables, and the same was then done with respect to the class II and class III criterion variables. Because these scale-criterion variable correlations were sometimes taken from two or more studies, averaging was also done in some instances across different samples. This was done in order to take best advantage of

TABLE 7-3 *(concluded)*

Class III criterion variables for the IMS scale:
- Clinical status
- ISS scores
- Marital problems—PSSP
- Personal sex life—PSSP
- Sex relationships—PSSP

Class III criterion variables for the ISS scale:
- Clinical status
- IMS scores
- Marital problems—PSSP
- Personal sex life—PSSP
- Sex relationships—PSSP

Class III criterion variables for the IPA scale:
- Clinical status
- IFR scores
- Parent-child relationships—PSSP
- Family relationships—PSSP

Class III criterion variables for the IFR scale:
- Clinical status
- IPA scores
- Parent-child relationships—PSSP
- Family relationships—PSSP

the available data from all the different studies that were conducted over the past seven years by the author and his colleagues. The average construct validity correlations are shown in Table 7-4.

The major observation that is to be made with respect to the correlations shown in Table 7-4 is that in all instances the class I average correlation was smaller than the average class II and class III correlations, and the average class II correlation was smaller than the average class III correlation. That is precisely the finding that must be obtained in order to support the claim that the CMP scales have construct validity. Again, it must be remembered that correlations do not represent an equal-interval metric, so the comparisons described above should actually be made by using the *squared* correlations. If that is done, the comparisons become even more dramatic, and the overall conclusion is that these data provide fairly strong evidence to support the claim that the CMP scales have good construct validity.

TABLE 7-4
Construct validity data for the CMP scales

CMP scales	Construct criterion variables		
	Class I	Class II	Class III
GCS	.11	.40	.68
ISE	.11	.36	.56
IMS	.11	.27	.64
ISS	.04	.24	.68
IPA	—*	.42	.76
IFR	.11	.15	.56

* These data were not available. Construct validity data for the CAM, CAF, and IPR scales are also not currently available.

FACTORIAL VALIDITY

Factorial validity, in the final analysis, is akin to both content and construct validity. Basically, one can define factorial validity, as used in this manual, as a form of item analysis. It is a way of examining the items on the CMP scales to determine whether they correlate with the things they are supposed to correlate with. More specifically, if an item on a particular scale is a valid measure of the construct that the scale is supposed to measure, the item should have a much higher correlation with that scale's total score than with the total score for any other scale.

Those who are familiar with "factor analysis" as a method of simultaneously analyzing the relationships among many variable or of generating "factors" may be tempted to think of factorial validity as emerging from a factor analysis of the items that make up the CMP scales. Such a view is correct, but because many fail to realize that there are many different types of factor analysis, there is a potential for confusion.

Factor analysis for the CMP scales

A great many people who work with the statistical analysis of social and behavioral science data think of factor analysis as consisting largely or exclusively of a procedure known as the principal components or PC method followed by a varimax rotation. Actually, that is only one of a large number of factoring techniques that are referred to as analytic or post hoc forms of factor analysis. Those who know about or use this or other types of analytic factor analysis may be tempted to use it in an effort to determine how many different factors are embedded within each of the CMP scales. *This should never be done*. The CMP scales are not multidimensional scales; they do not contain multifactor measures of the constructs they were designed to measure; and the development of such factors will constitute an abuse of the scales that will enormously threaten their primary utility and severely mislead those who attempt to use multiple factors that are derived from the scales.

Over the past several years I have conducted several of these kinds of factor analysis, and the same consistent results have always been obtained. These results are described here, without supporting evidence, in order to discourage an uninformed and misleading analysis of the CMP scale items. First, a proper and careful use of Cattell's (1966) SCREE test always shows that the CMP scales contain two factors. Second, these factors are not substantive, content, or construct factors. Third, they are method factors that arise as an artifact of using item reversals. One factor is heavily loaded by those items that are positively worded, and the other factor is defined by those items that are negatively worded. In short, analytic factor analysis is not capable of producing multiple construct factors for the simple reason that each of the CMP scales is a unidimensional measure of the construct or problem it was designed to measure. The only benefit that is derived from this type of factor analysis of the CMP scale items is to confirm the above results that suggest the use of item reversals helps to break up or minimize the effects of response set biases.

The greatest danger that attends the use of an analytic type of factor analysis arises from the fact that there are no good and accepted standards for deciding how many factors to extract. Cattell's (1966) SCREE test is one of the best criteria for deciding how many factors to extract and retain but it is a highly subjective procedure that can also be severely abused. The most popular approach is to extract and retain all factors with eigenvalues of 1.0 or greater but this criterion is often little more than a shot in the dark and it too can be quite misleading.

In spite of these cautionary remarks concerning the use of analytic forms of factor analysis, it is suspected that some will nonetheless use this strategy to examine the factor structure of the CMP scale items. Thus, while such a procedure is not recommended, some further cautionary advice is ventured. If an investigator insists on factor analyzing the items for any single CMP scale by using an analytic procedure, that should never be seriously undertaken with fewer than 500 observations. Second, the "eigenvalue = 1" criterion should never be used to determine how many factors to extract and retain. Cattell's (1966) SCREE test should be the method of choice in this regard and that should be used only when the investigator has acquired considerable experience in analyzing what Cattell and Dickman (1962) refer to as "plasmode" data. Finally, an attempt should not be made to use a common factor model by estimating and then analyzing "communalities." At this time no one knows what those might be, and no one has an acceptable data base or method for estimating them.

A second approach to the use of an analytic type of factor analysis with the CMP scales might be to factor all the items from two or more scales. Thus, if three scales were used that would require the factor analysis of 75 items; five scales would involve 125 items. If this approach is attempted, it should be based on a minimum of 20 observations for each item, and the number of factors that are extracted should be set equal to the number of scales that are used in the analysis. Such an approach as this represents an effort to use a "blind" analytic type of factor analysis as a means of achieving the results of an a priori or confirmatory factoring method. It is much wiser to use the proper method in the first place; a type of confirmatory or hypothesis-testing factor analysis. One such method is described next.

The multiple group method

If a measurement tool is constructed as a multi-item device, the use of those items implies the hypothesis that each item in some way measures the construct in question and not some other construct. This argument lies at the heart of the concepts of content and construct validity. However, the use of an item in a scale also implies the hypotheses that it will have a higher correlation with an independent measure of the construct, and it will have a lower correlation with other construct measures. The major task of investigating this kind of factorial validity consists of finding some way to test the hypotheses that items correlate well with the variables (constructs, factors) they are supposed to correlate with and that they correlate poorly with the constructs they are not supposed to correlate with. The multiple group method of factor analysis (Thurstone, 1947; Overall and Klett, 1972; Nunnally, 1978) is almost ideally suited for this task.

The multiple group method is a type of an a priori, confirmatory, or hypothesis-testing factor analysis that is designed to show whether some well-specified hypothesis matrix will do a good job of accounting for the pattern of correlations among a specific set of variables that are supposed to represent a specific set of well-defined factors. One of the niceties of the multiple group method consists of the fact that it is mathematically very simple. A more important nicety is that, when used to study the factorial validity of a measurement tool, it can be shown to be mathematically identical to a special form of item analysis and that the mathematics of factor analysis can be avoided entirely.

If R denotes the matrix of correlations among all the items for two or more scales (with units on the main diagonal), and H denotes an hypothesis matrix that specifies all the explicit hypotheses concerning which items load on which factors, then the factor loadings that are needed to confirm or deny those hypotheses can be obtained as

$$S = RHD^{-1/2} \quad (7.1)$$

where D is the diagonal of H′RH and

$$\phi = D^{-1/2}H'RHD^{-1/2} \quad (7.2)$$

contains the correlations among the factors. The great advantage of this method arises from the fact that the above results can be obtained by computing nothing more than simple Pearson product moment correlations. In order to compute the factor loading matrix, S, all that is needed is to compute a total score for each scale involved in the analysis and to then correlate every scale item with each of those total scores. The matrix, ϕ, is obtained by correlating all of the total scores with each other. This procedure is direct; it is powerful as an hypothesis-testing procedure; it requires no rotation; it produces an oblique solution; and it is simple to use.

The factorial validity findings

In most of the studies that were conducted to investigate the reliability and validity of the CMP scales, each client or respondent was asked to complete one or more scales in addition to the one that was being investigated. That was done in order to examine the factorial, content, and construct validity of the CMP scales at the item or indicant level as compared to the total-score or construct level. These investigations of the CMP factorial validity are not uniform from one study to the next because a different set of scales was used in each of them. Also, some of the factorial validity studies used variables other than the CMP scales. Even so, the same logic applies; the items comprising a scale should correlate better with that scale's total score than with any other total score or variable included in the analysis.

Before presenting the evidence concerning the factorial validity of the CMP scales, it should be recognized that the correlation of any scale item with its own total score is a part-whole correlation. It is a correlation between an item and the sum of 24 items and itself. The presence of this item-self correlation could, in some cases, present a somewhat inflated picture of the factorial validity of the CMP scales. However, that risk was removed by adjusting all item-total correlations to remove the unwanted item-self correlation, and the procedure for doing that can be found in the text by Nunnally (1978).

The GCS, ISE, IMS, and ISS scales. The best evidence concerning the factorial validity for the GCS, ISE, IMS, and ISS scales was obtained by combining the samples from several different studies that used each of them. One study concerned the relationships between these scales and the family lifecycle model (Murphy, Hudson, and Cheung, 1980); another concerned their relationships to sexual activity and preferences (Nurius, 1982; Hudson and Nurius, 1981); and the third concerns a study currently in progress to examine their relationships to spouse abuse (Hontanosas, Cruz, Kaneshiro, and Sanchez, 1979; Hudson, Harrison, and Maxwell, 1981; Hudson and McIntosh, 1981). By combining these studies, the overall sample that was used to investigate the factorial validity of these scales consisted of 1,161 responses. For a few of the bivariate relationships that emerged from this combined sample, there were as few as 792 cases but the overwhelming majority was based on 1,100 or more cases.

Each of the items for the GCS, ISE, IMS, and ISS scales was correlated with their total scores, the total score for the SAS scale (Hudson, Murphy, and Nurius, 1981), and with the respondents' age, sex, income, and number of years of school completed. The factorial validity analysis for this set of data produced a single table of item-total correlations that contains 125 rows and nine columns. However, in order to make a more orderly presentation of these findings, these item-total correlations are broken up and presented as four separate tables. Also, the item-total correlations for the SAS items are omitted since the SAS scale is not part of the CMP.

The item-total correlations that provide information concerning the factorial validity of the GCS scale are shown as Table 7-5. Examination of those correlations shows that all but one of the GCS items had an equal or higher correlation with the GCS total score than with any of the other eight measures. Item 20, which was originally written as "I feel that I don't deserve to have a good time," correlated more highly with the ISE total score than it did with the GCS total score. At first this finding was suprising because it is well known that clinically depressed people are not usually "fun loving" individuals.

In light of this finding, it was presumed that item 20 was weak because it measured the feeling that one does not "deserve" to have a good time. On a post hoc basis it was reasoned that depressed clients

TABLE 7-5
GCS factorial validity data

GCS item	GCS	ISE	IMS	ISS	SAS	AGE	SEX	SCH	INC
1	.53	.46	.28	.27	−.01	−.23	.06	−.01	.11
2	.57	.45	.28	.21	−.11	−.31	.11	.03	.18
3	.31	.24	.23	.19	−.10	−.33	.01	−.03	.19
4	.38	.27	.20	.16	−.11	−.21	.25	.07	.07
5	.35	.25	.16	.18	.06	.01	.03	−.03	.02
6	.34	.31	.15	.13	−.13	−.31	.06	.02	.20
7	.28	.22	.11	.14	.12	.06	.01	−.09	−.02
8	.40	.29	.28	.28	.10	.03	−.18	−.14	−.03
9	.60	.46	.30	.33	.09	−.06	−.01	−.11	.03
10	.64	.50	.31	.29	−.06	−.23	.10	.01	.14
11	.48	.42	.26	.28	.08	−.07	−.09	−.13	−.01
12	.57	.56	.32	.35	.08	−.08	−.08	−.08	.14
13	.46	.36	.14	.21	.04	−.03	−.08	−.06	−.01
14	.47	.47	.18	.26	.14	−.07	−.04	−.16	.02
15	.35	.33	.10	.12	.06	.07	−.11	−.01	−.05
16	.50	.45	.20	.22	−.02	−.22	.15	−.06	.15
17	.57	.47	.28	.30	.07	−.14	.04	−.11	.07
18	.38	.34	.20	.22	−.01	−.15	.13	−.02	.07
19	.42	.36	.17	.22	.08	−.08	.13	−.09	.06
20	.32	.33	.18	.25	.16	.02	−.02	−.11	.02
21	.57	.47	.37	.32	−.03	−.23	−.01	−.08	.11
22	.47	.42	.28	.27	.11	−.01	−.15	−.09	.01
23	.61	.53	.36	.36	−.02	−.06	−.01	.01	.03
24	.49	.38	.30	.27	−.05	−.25	.04	.04	.16
25	.51	.43	.24	.26	.11	−.10	.03	−.08	.05

may feel they do *deserve* to have a good time but that for some reason they just do not. The item was therefore changed to read, "It is hard for me to have a good time." Although no data are available to evaluate the validity of item 20 as it is presented in the present version of the GCS, it is believed that this change will prove to be a small improvement in the scale.

When evaluating the GCS item validity data, it should be remembered that depression and problems with self-esteem tend to be rather highly correlated in the general population. It is therefore to be expected that several of the GCS items will be at least moderately correlated with the ISE total score and that is clearly revealed in Table 7-5. The presence of the ISE scale, therefore, provides a critical test of the item content, construct, and factorial validity for the GCS scale. The major finding, however, is that all but one of the GCS items correlated as well or better with the GCS total score than with the ISE total score. Item 14 correlated at .47 with both the GCS and ISE total scores, but this fact and the size of the correlation merely indicate that the item taps a significant component of both depression and self-esteem problems.

The factorial validity data for the ISE scale are shown in Table 7-6 where it can be seen that each of the ISE items correlated more highly with the ISE total score than with any of the other eight measures. As with the GCS items, many of the ISE items have moderate correlations with the GCS scale as well. The presence of the GCS scale in these data can be seen as providing a critical test of the item content, construct, and factorial validity for the ISE scale because these two measures tend to be highly correlated in the general population.

The factorial validity data for the IMS scale are shown in Table 7-7, and it should be recognized that because marital and sexual discord tend to be highly correlated in the general population, the presence of the ISS scale in these data provides a critical test of the IMS item content, construct, and factorial validity. Examination of the data shown in Table 7-7 shows that all but two of the IMS items have higher correlations with the IMS total score than with any of the other eight measures. Item 21 was originally worded to read, "I feel that my partner is pleased with me as a sex partner." It is therefore no surprise that it has a higher correlation with the ISS scale

TABLE 7-6
ISE factorial validity data

ISE item	GCS	ISE	IMS	ISS	SAS	AGE	SEX	SCH	INC
1	.34	.37	.18	.21	.10	−.09	−.03	−.07	.02
2	.56	.62	.28	.28	.11	−.14	−.01	−.15	.05
3	.42	.53	.20	.27	.18	.07	−.02	−.10	−.02
4	.54	.69	.30	.31	.10	−.09	−.08	−.05	.07
5	.52	.66	.23	.28	.14	−.08	−.12	−.09	.04
6	.51	.59	.26	.33	.12	−.04	.02	−.11	.02
7	.54	.69	.26	.30	.12	−.08	.11	−.10	.05
8	.45	.54	.22	.22	.02	−.25	.16	−.09	.11
9	.40	.52	.23	.22	.01	−.36	.11	−.08	.20
10	.51	.68	.25	.31	.10	−.11	.02	−.11	.03
11	.48	.54	.28	.28	.04	−.20	.11	−.11	.07
12	.55	.61	.30	.35	.08	−.14	.03	−.06	.06
13	.48	.65	.25	.30	.08	−.05	−.01	−.06	.01
14	.46	.64	.23	.29	.21	.02	−.07	−.14	.03
15	.41	.50	.17	.26	.15	.01	−.01	−.10	.01
16	.42	.55	.21	.21	.07	−.22	.07	−.11	.11
17	.47	.58	.25	.32	.20	−.10	.04	−.21	.10
18	.53	.70	.28	.35	.15	−.02	−.03	−.08	.02
19	.44	.56	.21	.24	.03	−.09	.03	−.08	.03
20	.42	.48	.26	.25	.08	−.13	.12	−.07	.04
21	.43	.56	.23	.27	.19	.01	−.04	−.14	.08
22	.54	.71	.28	.30	.11	−.07	−.04	−.07	.06
23	.48	.64	.25	.28	.15	−.01	−.06	−.10	.03
24	.45	.60	.27	.25	−.02	−.26	.07	−.03	.10
25	.48	.63	.29	.33	.13	−.09	−.03	−.09	.04

TABLE 7-7
IMS factorial validity data

IMS item	GCS	ISE	IMS	ISS	SAS	AGE	SEX	SCH	INC
1	.33	.30	.61	.52	.02	−.05	.05	−.05	.03
2	.28	.24	.54	.39	.03	−.05	.01	−.03	.04
3	.32	.27	.62	.44	.07	−.05	.01	−.07	.02
4	.22	.16	.63	.39	−.03	−.11	.09	.01	.04
5	.24	.19	.54	.37	.02	−.07	.08	−.05	.03
6	.30	.23	.73	.46	−.13	−.25	.06	.01	.12
7	.33	.26	.65	.45	−.01	−.15	.08	−.06	.04
8	.34	.27	.81	.57	−.06	−.16	.08	.01	.09
9	.37	.31	.84	.59	−.09	−.15	.09	.04	.10
10	.34	.30	.64	.54	−.03	−.06	.05	.02	.02
11	.31	.30	.72	.57	−.02	.01	.05	.03	.01
12	.32	.27	.52	.38	−.01	−.11	.10	−.04	.01
13	.32	.29	.78	.52	.01	−.12	.12	.01	.06
14	.21	.18	.52	.34	.03	−.05	.09	−.03	−.04
15	.27	.25	.60	.47	−.05	−.07	.07	.02	−.01
16	.35	.34	.64	.43	−.04	−.16	.07	−.01	.07
17	.31	.27	.39	.27	−.11	−.29	.08	−.02	.15
18	.22	.15	.73	.46	−.03	−.13	.12	−.01	.02
19	.33	.30	.83	.56	−.01	−.10	.05	−.01	.04
20	.38	.30	.80	.51	−.12	−.25	.08	.01	.12
21	.33	.30	.54	.67	−.02	−.04	−.03	.04	.02
22	.22	.24	.21	.22	.16	−.01	−.13	−.19	.04
23	.34	.27	.78	.52	−.13	−.25	.10	.06	.10
24	.28	.24	.76	.57	−.01	−.07	.10	−.01	−.01
25	.27	.24	.61	.58	−.03	−.03	.05	.01	.01

than with the IMS scale. From a content and construct validity point of view, it was a mistake to have put this item into the IMS scale. The item was therefore replaced with the item which reads, "I feel that my partner is a comfort to me." Item 22 was originally worded to read, "I feel that we should do more things together." According to the data shown in Table 7-7, this item did not correlate very well with the IMS total score, and it had about the same correlation with the ISS scale. This finding seems to suggest that "doing more things together" does not provide a good indicator of the status of the relationship. This item was therefore replaced by the item that reads, "I feel that I no longer care for my partner."

Because these two items on the IMS scale were changed after the validation studies were completed, and on the basis of findings obtained from them, no data are currently available to examine their validity. Nonetheless, it is felt these changes will produce improvement in the IMS scale. The essential finding is that the other IMS items appear to have excellent factorial validity.

The ISS factorial validity data are shown in Table 7-8 where it can be seen that the great majority of the ISS items correlate more highly with the ISS total score than with any of the other eight measures. However, there were five items on this scale that required replacement. Item 4 appears to do a better job of measuring marital discord, and a careful inspection of the content of this item suggests the reason is that it serves more as a complaint about the marriage than about the sexual relationship. Item 14 appears to do a better job of measuring sexual *attitude* than discord, and item 16 correlates equally well with the ISS, IMS, GCS, and ISE total scores. Item 20 correlates more highly with the SAS total score than with the ISS total score, and it has a rather high correlation with the IMS total score. Finally, item 24 has a higher correlation with the IMS scale than it does with the ISS scale. These five items were originally written:

4. I feel that my partner sees little in me except for the sex I can give.
14. I feel that sex is something that has to be endured in our relationship.
16. My partner observes good personal hygiene.

TABLE 7-8
ISS factorial validity data

ISS item	GCS	ISE	IMS	ISS	SAS	AGE	SEX	SCH	INC
1	.33	.33	.50	.65	.06	−.02	−.02	−.03	−.02
2	.33	.33	.56	.72	.05	−.04	.09	.06	.01
3	.36	.32	.57	.77	.12	.03	.05	−.06	−.04
4	.34	.30	.47	.39	.14	.01	.05	−.14	−.03
5	.25	.32	.23	.38	.18	−.10	.05	−.07	.03
6	.27	.23	.48	.62	.01	−.07	.05	.01	.04
7	.26	.26	.35	.50	.14	−.04	.04	−.13	−.03
8	.29	.31	.55	.71	.06	−.03	−.01	−.02	−.01
9	.26	.25	.52	.71	.12	.09	.01	−.01	−.05
10	.33	.31	.46	.72	.17	.01	−.01	−.07	−.01
11	.26	.31	.33	.42	.14	−.07	.11	−.11	−.03
12	.31	.35	.33	.55	.26	−.01	.12	−.10	−.01
13	.23	.26	.35	.38	.12	−.08	.15	−.05	−.02
14	.14	.13	.08	.19	.32	.06	−.11	−.29	−.01
15	.22	.20	.26	.35	.09	−.04	.11	−.05	.03
16	.24	.20	.21	.24	.08	−.06	.01	−.10	.05
17	.35	.30	.38	.54	.15	−.12	.05	−.12	.03
18	.15	.18	.26	.33	.05	.03	−.18	−.07	.01
19	.31	.26	.48	.61	.08	−.19	.05	−.01	.01
20	.14	.06	.44	.28	−.31	−.21	−.22	.10	.13
21	.29	.26	.45	.67	.15	.01	.14	−.05	−.03
22	.32	.33	.46	.68	.08	.01	−.03	−.02	−.01
23	.31	.25	.45	.65	.10	.03	−.05	−.07	−.04
24	.16	.09	.28	.23	−.09	−.05	−.19	.01	.05
25	.26	.25	.53	.68	.03	−.01	.05	.03	.02

20. I would like to have sexual contact with someone other than my partner.
24. I feel that I should have sex more often.

In an effort to improve the quality of the ISS scale, these items were replaced with the following.

4. Sex with my partner has become a chore for me.
14. I try to avoid sexual contact with my partner.
16. My partner is a wonderful sex mate.
20. My partner seems to avoid sexual contact with me.
24. My partner does not satisfy me sexually.

As with the changes made for the GCS and IMS scales, there are no data currently available to assess the factorial validity of these new ISS scale items. It is believed, however, that they will provide a modest improvement in this scale. The major finding with respect to the data shown in Table 7-8 is that the great majority of the ISS items appear to have very good factorial validity.

The IPA scale. The factorial validity data for the IPA scale were obtained from a study of 93 persons who were seeking counseling services for one or more personal or interpersonal relationship problems (Hudson, Wung, and Borges, 1980). In addition to completing the IPA scale, each client was asked to also complete the ISE and ISS scales. Since this study was conducted largely to assess the known-groups discriminant validity of the IPA scale, the factorial validity of the IPA items was investigated by correlating each item with the clinical group status of the clients and with the total scores for the IPA, ISE, and ISS scales, and the data are shown in Table 7-9. There it can be seen that virtually all of the IPA items had higher correlations with the IPA total score than with either of the other two scales. The correlations with clinical group status serve primarily to confirm that each of the IPA items makes a large and significant contribution to the overall discriminant validity of the IPA scale. On the basis of these findings, it was concluded that the IPA has excellent factorial validity.

The CAM and CAF scales. The factorial validity data for the CAM and CAF scales were obtained from a survey of youths who were attending a middle-class boarding school in a large metropolitan area (Hudson, Hess, and Matayoshi, 1979). The sample consisted of 408 responses from youths attending grades 9 through 12. Each student completed the CAM and CAF scales but none of the other scales from the CMP. However, school officials provided access to

TABLE 7-9
IPA factorial validity data

IPA item	IPA	ISE	ISS	CRIT
1	.75	.38	.46	.78
2	.83	.44	.43	.77
3	.79	.38	.64	.77
4	.67	.23	.22	.66
5	.82	.49	.47	.76
6	.65	.28	.22	.73
7	.73	.31	.41	.62
8	.86	.53	.60	.77
9	.69	.39	.36	.76
10	.70	.36	.43	.59
11	.60	.30	.57	.58
12	.84	.45	.49	.81
13	.41	.17	.15	.39
14	.55	.36	.12	.53
15	.77	.34	.33	.71
16	.73	.31	.36	.75
17	.45	.25	.25	.43
18	.86	.46	.51	.82
19	.76	.43	.41	.75
20	.58	.26	.25	.55
21	.74	.41	.48	.68
22	.84	.47	.46	.79
23	.75	.44	.38	.74
24	.83	.39	.46	.73
25	.78	.40	.42	.64

the students' grades and their scores on the National Educational Development Test or NEDT. The factorial validity of the CAM and CAF scales therefore consists of the correlations between these scale items and the total scores for the CAM, CAF, NEDT, and cumulative grade-point average or GPA.

The factorial validity data for the CAM and CAF scales are shown in Tables 7-10 and 7-11 where it can be seen that all of the items for these two scales had higher correlations with their respective total scale scores than with any of the other three measures shown in the tables. On the basis of these findings, it was concluded that the CAM and CAF scales appear to have good factorial validity.

The IFR Scale. The reliability and validity of the IFR scale were investigated in a clinical study of 120 persons who were seeking help with one or more personal or interpersonal relationship problems (Hudson, Acklin, and Bartosh, 1980). Each respondent completed the IFR, GCS, Zung, and Beck scales as part of the research questionnaire. The Zung and Beck scales are measures of depression (Zung, 1965; Beck, 1961) and these were included in order to use the data for partial validation of the GCS scale in a separate study (Hud-

TABLE 7-10
CAM factorial validity data

CAM item	CAM	CAF	GPA	NEDT
1	.68	.34	−.14	−.06
2	.72	.28	−.15	−.08
3	.63	.33	−.07	.06
4	.71	.30	−.12	−.09
5	.46	.19	−.02	−.04
6	.55	.25	−.11	−.01
7	.62	.22	−.11	−.01
8	.70	.32	−.07	−.04
9	.42	.20	−.11	−.08
10	.50	.33	−.13	−.11
11	.49	.24	−.11	−.13
12	.71	.37	−.13	−.02
13	.49	.19	−.18	−.08
14	.47	.28	−.13	−.06
15	.74	.35	−.11	.01
16	.62	.34	−.17	−.01
17	.59	.29	−.08	−.09
18	.68	.34	−.08	.01
19	.61	.34	−.14	−.03
20	.44	.19	−.09	−.07
21	.59	.27	−.09	−.05
22	.39	.26	−.13	−.08
23	.58	.40	−.08	−.05
24	.58	.29	−.09	−.09
25	.60	.27	−.06	−.14

TABLE 7-11
CAF factorial validity data

CAF item	CAF	CAM	GPA	NEDT
1	.73	.27	−.09	−.12
2	.77	.36	−.10	−.11
3	.67	.34	−.02	−.12
4	.69	.38	−.09	−.13
5	.58	.28	−.10	−.14
6	.51	.26	−.15	−.25
7	.66	.26	−.04	−.12
8	.74	.40	−.13	−.07
9	.55	.31	−.11	−.24
10	.58	.36	−.11	−.16
11	.63	.34	−.08	−.18
12	.79	.43	−.07	−.09
13	.68	.30	−.07	−.14
14	.61	.28	−.05	−.06
15	.79	.42	−.06	−.11
16	.66	.33	−.08	−.05
17	.65	.33	−.01	−.14
18	.70	.31	−.10	−.13
19	.71	.32	−.12	−.19
20	.53	.24	−.11	−.24
21	.59	.31	−.06	−.13
22	.50	.20	−.04	−.09
23	.68	.39	−.07	−.14
24	.53	.26	−.07	−.20
25	.63	.28	−.04	−.10

son, Hamada, Keech, and Harlan, 1980). The factorial validity of the IFR scale was investigated by correlating each of its items with the total scores for the IFR, GCS, Zung, and Beck scales and with the clinical status of the clients (CRIT). The correlations with the clinical criterion group status were used to determine whether the IFR items would do a good job of discriminating between the groups, and the findings are shown in Table 7-12. Examination of the data shown in Table 7-12 strongly suggests that the IFR has excellent factorial validity since the items have higher correlations with the IFR total score than with any of the other measures shown in the table.

SUMMARY COMMENTS

At the beginning of this chapter it was said that a final comment about the content validity of the scales would be reserved until the remaining evidence concerning the validity of the CMP scales had been presented. The main point to be made at this time is that the distinctions that are made between content, construct, and factorial validity are in part conceptual and in part are made on methodologi-

TABLE 7-12
IFR factorial validity data

IFR item	IFR	GCS	Zung	Beck	CRIT
1	.79	.57	.62	.47	.74
2	.87	.56	.53	.40	.64
3	.82	.70	.72	.59	.66
4	.86	.61	.58	.51	.62
5	.79	.67	.60	.51	.58
6	.80	.65	.60	.55	.53
7	.80	.73	.65	.61	.66
8	.82	.61	.57	.53	.63
9	.82	.64	.63	.55	.68
10	.81	.59	.56	.45	.68
11	.85	.65	.63	.60	.74
12	.82	.63	.63	.56	.69
13	.83	.75	.76	.61	.67
14	.82	.62	.56	.48	.61
15	.61	.55	.50	.43	.40
16	.83	.63	.70	.55	.67
17	.88	.60	.60	.49	.71
18	.83	.53	.60	.41	.62
19	.86	.62	.65	.52	.68
20	.91	.69	.63	.56	.66
21	.80	.61	.57	.49	.54
22	.86	.66	.66	.56	.71
23	.88	.63	.69	.54	.61
24	.82	.68	.61	.54	.68
25	.85	.72	.72	.62	.67

cal grounds. However, it should be realized that they all tend to mutually support one another, and in the final analysis some of the distinctions tend to be seen as a bit artificial. For example, the data concerning the factorial validity of the CMP scales have a great deal to say about their construct validity for the simple reason that the data show at the item level of analysis that the scales tend very much to correlate with measures they are supposed to correlate with and that they correlate rather poorly (or less strongly) with those measures they are theoretically unrelated to. In other words, the criteria that were established to examine the factorial validity of the scales conform very well to the definitions of convergent and discriminant validity that were stated by Campbell and Fiske (1959).

The same argument can be made with respect to the known-groups concurrent (discriminant) validity data. Again, the CMP scales showed a marked tendency to do a better job of correlating with the criterion group status than did other measures and the criteria that were used to examine this type of validity closely parallels the definition of construct validity that was used in all the studies. Finally, the factorial validity data provide considerable support for the content validity of the scales simply because the items on each were shown to correlate very well with their overall total scale scores. The major support for this claim arises from the fact that items are predicted to correlate with their respective total scores on the basis of their contents—and nothing more. This was clearly illustrated by the mistake that was made in the development of the IMS scale. One item was assigned to it without regard to content, and content alone indicated it belonged with another scale.

As was said at the beginning of this chapter, the data reported here constitute the most important findings currently available with respect to the performance of the CMP scales. It is believed that the findings reported in this chapter provide reasonable evidence to support the claim that the CMP scales actually measure the problems or constructs they were designed to measure; on the basis of these findings, they can be regarded as valid measures for use in monitoring and assessing certain client problems in clinical practice.

There are obvious gaps in the available information about the CMP scale validities, but some of those will be filled by studies that are currently in progress and by others that will no doubt be conducted in the future. As further data are made available, it may be found that one or more of the scales can be improved by making other modest revisions of item contents. Until such data are available, however, it appears that the scales will do a credible job in their present form and structure.

8 Cutting scores, false positives, and false negatives

One of the most attractive features of the CMP scales arises from the fact that each one has a clinical cutting score, and the same cutting score can be used for each of the scales. These features are especially attractive because clinicians and researchers can use the cutting score as a diagnostic or therapeutic criterion or benchmark, and they do not have to remember a different cutting score for each of the nine scales. This chapter presents a discussion of the CMP cutting scores, describes their use in clinical practice and in research, and discusses the methods that were used to establish a cutting score for each of the scales. Also the available data are presented regarding rates of false positives and false negatives.

CUTTING SCORES

At this point the reader will be thoroughly familiar with the fact that each of the CMP scales measures the degree or severity of a specific personal or interpersonal relationship dysfunction. It has been said on several occasions that large scores on the CMP scales represent more serious problems, and lower scores indicate the relative absence of such problems. However, one may introduce the question as to how large is "large" and how small is "small"?

This is an important issue for those who are seeking measurement devices that will aid the clinician's diagnostic work and those who would like to establish therapeutic goals in terms of the measured client problem. Because of the structure of the CMP scales, it is apparent that one should be virtually unconcerned about a client who produces a score of zero on any of the CMP scales, provided the client has given accurate and candid responses. What about a score of 5, 10, 15, or even 20?

From any reasonable clinical point of view, it is patently unrealistic to expect that everyone who scores above zero is a good candidate for the use of clinical services. Such services are expensive; personnel who provide them are not abundantly available; and time that is devoted to clients who have only very small or trivial problems is usually regarded as a costly luxury.

It is equally unrealistic to expect that anyone will always be completely free of any of the problems that are measured by the CMP scales (exceptions would consist of those for whom the scales are not relevant—childless individuals will not have parent-child relationship disorders). All of us experience the slings and arrows of the ill-fortune normally associated with adult living (including therapists and researchers!), so scores that are obtained from use of the CMP scales with any population almost always will show that most people score above zero on the CMP scales. In short, most of us suffer to some degree with respect to all of the problems that are measured by the CMP scales and that are relevant to our personal situations and circumstances.

From a clinical point of view, the issue is not whether we are completely free of any personal or social dysfunction but whether the severity or magnitude of a problem reaches a level that it can be regarded as being clinically significant. It is conceivable that some specific measurement tool might be capable of detecting the presence of a clinically significant problem in personal and social functioning but that the score demarking such a problem would be unique for the individual. In such a case, it would not be possible to define a general cutting score to be used as an indicator of the presence of a clinically significant disorder because such a cutting score would be different from one person to the next.

In order to develop a useful cutting score for clinical and research applications, it is necessary that the measurement scale perform in a manner that yields the same, or approximately the same, cutting score for a very large proportion of those who will complete the scale. If a scale does perform in such a manner, it should be possible to locate a score along the measurement continuum that effectively divides or separates individuals, samples, or populations into at least two groups: those who have a clinically significant problem in the area being assessed and those who do not have such a problem. In other words, a clinical cutting score is defined here as that score on any of the CMP scales above which indicates the individual has a clinically significant problem and below which indicates the individual has no such problem.

USES FOR CUTTING SCORES

The availability of an effective cutting score for the CMP scales provides for two important uses in clinical practice. One relates to diagnosis and the other to the setting of therapeutic goals. When discussing the use of the CMP scales as diagnostic aids, it is important to distinguish between two different types of diagnostic activity. One type of diagnosis consists of an effort to identify the client as a member of some nosological group or to categorize the client within a nosological typology. An example of this would be an effort to describe a depressed client as having some particular *type* of depression, that is, psychogenic, situational, endogenous, involutional, and so forth. Another example would consist of an effort to classify a psychotic patient according to the *type* of psychosis, that is, manic-depressive, paranoid, schizophrenic, and so forth. It should be carefully noted that none of the CMP scales was designed for use in such diagnostic efforts; it is fully expected that they would perform very poorly in this regard, and they should never be used in such a manner.

The second type of diagnostic activity consists of an effort to determine whether a client has a clinically significant problem with respect to some specific area of personal or social functioning. This is a different and much simpler form of diagnostic activity and it is one for which the CMP scales are well suited. For example, if it is presumed that most or all clients will have some degree of depression (few are ever completely free of depression), the task is one of determining whether the amount of depression is great enough to be regarded as clinically significant.

If the amount of depression that is exhibited or reported by a client is not clinically significant, there is no reasonable basis for instituting a program of treatment that is designed to reduce depression. However, if a client's problem is large enough to be regarded as clinically significant, the therapist would surely want to know that and then consider whether that specific problem should be regarded as the primary target for treatment. This sort of reasoning strongly suggests that prothetic diagnoses (those based on an assessment of the severity or magnitude of a disorder), as compared to metathetic diagnoses (those based on an assessment of the types of disorder), may be conceptually much simpler but comprise an extremely important diagnostic activity. In the final analysis, it is probably much more important than a metathetic diagnosis.[1] The availability of an effective cutting score for the CMP scales therefore enables them to serve as important prothetic diagnostic aids in determining whether the client has a clinically significant problem in personal or social functioning.

The second clinical use of an effective cutting score for the CMP scales arises in connection with the setting of specific therapeutic goals for a specific client and the assessment of the final treatment outcome to determine whether the therapeutic goals were actually achieved. In order to illustrate this use of the CMP cutting score, consider the case of a 45-year-old man who complains of difficulties in his relationship with his wife and some difficulty in working with people at his job. Suppose a therapist administered the IMS and IPR scales to such a client and obtained scores of 64 and 47, respectively. Since the cutting score for the CMP scales is 30, these obtained scores indicate that the client has a clinically significant problem in both areas.

In some cases, the prognosis is quite good, and the therapist might very well establish a treatment goal of completely eliminating both the marital and the peer relationship disorder. If that were the case, the clinical cutting score becomes a very important criterion against which the therapist can evaluate the final outcome of the treatment effort. It also serves as a therapeutic goal, and the attempt is clearly one of reducing the magnitude of both problems to the point where the client will obtain IMS and IPR scores below 30.

Actually, it is presumptuous to hope that we can ever completely eliminate any specific personal or social dysfunction of the sort that is measured by the CMP scales. The best one should hope for is to reduce them to the point where the client obtains scores that are no longer in the clinically significant range, that is, below 30. But how far below 30? As indicated earlier, that is a decision that must be negotiated in most cases with the client. However, one guideline in this regard can be taken from the earlier discussions that concern the SEM associated with each of the scales. Since the cutting score is an imperfect indicator of the presence of a clinically significant problem and this arises primarily because of the imperfect reliabilities of the scales, it may be wise to use the SEM to establish a lower bound in deciding whether further treatment would be useful. That is, since the SEM for the scales is usually about five points, it may be wise to terminate treatment for a specific problem once the client scores

---

[1] The use of *prothetic* and *metathetic* to describe these two fundamentally different kinds of diagnostic activities is entirely consistent with, and based upon, two important conceptual distinctions provided by Stevens (1968).

consistently below 25. In other words, if a client consistently scores below 25 on any of the CMP scales, that constitutes fairly strong evidence of the absence of a clinically significant problem in the area being assessed. If the client does not have a clinically significant problem, there is little justification for continuing a treatment relationship.

Although the prognosis will be quite good for some clients, the experienced therapist knows that is certainly not the case for others. In such cases it may be wholly unrealistic to hope that a short or even intermediate period of treatment will produce changes great enough to expect the client to score below 30 (much less 25) on a specific scale. In a number of cases such an expectation would not be realistic even after very long periods of continuous therapy.

When the prognosis for therapeutic gain is clearly very poor, the therapist should not use the clinical cutting score as a well-defined therapeutic goal. This does not mean that such a goal is abandoned entirely. It is well known that goal achievement is best obtained in connection with goal setting. If goals are not set, the likelihood of their achievement is diminished—often by a considerable margin. Nonetheless, it must also be recognized that for cases in which the prognosis is very poor, the setting of unrealistic goals may also work to the ultimate disadvantage of the client. What, then, might be used as some form of guideline for therapeutic goal setting with such cases? Again, the SEM is an excellent candidate in this regard.

Suppose, for example, that a therapist is confronted with a family in which there are multiple problems and in which the relationship between the mother and one of two children is severely deteriorated. In such a case one might obtain from a child a CAM score of 87 and the mother might produce an IPA score of 93. Given the severe problems in this parent-child relationship, the therapist might well consider the modest goal of attempting to produce *some* significant gain—however small. In short, the primary goal might be one of trying to get the relationship moving in the right direction. If that were the case, the therapist might feel grateful for a gain of about five points for the child or the mother or both.

Once some positive movement is achieved in cases such as this, the next step is to set another equally modest goal; perhaps another five-point gain. In other words, goal setting in such cases might have to be an incremental process over an extended period of time. This does not necessarily mean that the incremental targeting for change must remain a fixed process or rule. If large gains are eventually made through such incremental goal setting, it may later become possible (reasonable) to view the clinical cutting score as a realistic therapeutic goal. This is not to suggest that such will be the outcome in all or even most of the very difficult cases that therapists must deal with. It is to suggest that the process of goal setting is extremely important and that it is usually a good procedure to state reasonable goals as part of the case record.

While the above represent the most important uses of the clinical cutting scores for the CMP scales, it should be noted that they also have at least one important research application. When one is using the CMP scales in comparative and survey studies, the availability of the clinical cutting scores provides an excellent basis for examining the incidence and prevalence of the personal and social dysfunctions that are measured by the scales (Murphy, Hudson, and Cheung, 1980). For experimental studies the cutting scores provide a basis for determining whether certain experimental subjects should be excluded from study samples. For example, if one were testing an experimental intervention to determine whether it would significantly reduce, say, the severity of self-esteem problems, it would be important to verify that experimental subjects indeed had a clinically significant problem with self-esteem. Persons who score below 30 would likely be good candidates for exclusion from the study, and one might even choose to eliminate those who scored below 35. How can an intervention affect a problem that isn't there?

THE CUTTING SCORE METHODOLOGY

From the beginning of this project, it was decided that a clear methodology had to be developed for establishing an appropriate cutting score for each of the CMP scales. It was fully expected that the cutting score for one scale would not be the same for another nor was it expected that they would even be very similar to each other. Although it turned out to be the case that all the cutting scores appear to be the same, it can only be regarded as a fortunate happenstance that will be discussed at the end of this chapter. In the meantime, it is important to describe the methods that were developed for establishing the clinical cutting scores for the scales.

The basic methodology for establishing the cutting scores for the CMP scales is intimately linked with the method that was used for examining the known groups form of concurrent criterion validity. That is, the clinicians' judgments regarding the presence or absence of a serious problem in a specific area of personal or social functioning was used as the fundamental basis for establishing a clinical cutting score. In each of the validation studies, the clinician independently classified each client as either having a serious problem in the area being assessed or as being free of such a problem. Once that was done, the client completed the appropriate CMP scale, and a separate cumulative frequency distribution for the scale scores was prepared for the two criterion groups. The two cumulative frequency distributions were then used to locate the particular scale score which, if used as the cutting score, would minimize the sum of the false positives and false negatives. That score was chosen as the optimal cutting score.

In some of the later validation studies, it became apparent that exclusive reliance on the clinicians' judgments carried with it the implicit assumption that their decisions in classifying the clients were made with perfect reliability. Such an assumption is untenable, and an effort was made to improve the reliability of the criterion measure. It was decided that only those cases would be retained in which both the client and the therapist agreed on the problem status. It was believed this procedure would reduce some of the ambiguity or uncertainty that is involved in obtaining clinician judgements with regard to the presence of clinically significant problems.

Some might object to this procedure under the claim that clients are rarely, if ever, in a good position to make or participate in a decision concerning whether theirs is a clinically significant problem. Others might object under the claim or belief that the therapist is always the best judge of the matter. Nonetheless, the procedure was used because it is thought we should never be so arrogant as to believe that we always know more about our clients' problems than they know. Neither should we be so humble as to believe that clients always know more than we can know about the presence of clinically significant personal and social dysfunctions. In the final analysis, there are only two ways to determine whether a client has a problem: watch them or ask them (Hudson, 1981b). If the client states there is a problem, but the therapist disagrees, how is the dispute resolved? The same question arises when the therapist states there is a problem but the client disagrees. Similar difficulties arise when one actor claims there is no problem and the other disagrees. For purposes of examining the clinical cutting scores for the CMP scales, it was thought that the risk of criterion error was significantly reduced when only those cases were retained in which both the therapist and the client were in agreement as to the presence or absence of a clinically significant problem in any of the areas that were being assessed. While the procedure had the effect of reducing sample sizes, it also had the very desirable effect of reducing the rates of false positives and false negatives.

CUTTING SCORE FINDINGS

Details concerning the cutting score findings can be summarized for all the scales and those data are shown in Table 8–1. For several

TABLE 8-1
Cutting scores and classification error rates

Scale	N	Cutting score	False positives	False negatives	Overall error
GCS	145*	31	14.2%	13.0%	13.8%
ISE	124	30	8.3%	25.0%	17.7%
IMS	96	26	8.0%	8.8%	8.4%
		(30)	(6.9%)	(11.0%)	(8.5%)
ISS	100	28	11.8%	14.3%	13.0%
		(30)	(7.8%)	(20.4%)	(14.0%)
IPA	62*	24	2.6%	0.0%	1.6%
		(30)	(2.6%)	(8.7%)	(4.8%)
CAM	38	33	n.a.	n.a.	n.a.
CAF	30	33	n.a.	n.a.	n.a.
IFR	87*	33	0.0%	2.5%	1.2%
		(30)	(4.3%)	(2.5%)	(3.4%)
IPR	n.a.	n.a.	n.a.	n.a.	n.a.
Mean		29.8	7.5%	10.6%	9.3%
		(30)	(7.2%)	(13.4%)	(10.4%)

n.a. = not available.
* This sample consists of only those cases in which both the therapist and client agreed with respect to the presence or absence of a problem.

of the scales there are two sets of information provided. The first line in the table for those scales reports the findings for the best or optimal cutting score and the second line reports the findings when a score of 30 is used as the clinical cutting score.

Throughout this manual the reader has been advised that the score of 30 serves as the appropriate cutting score for each of the scales. However, examination of the data shown in Table 8-1 shows that the optimal cutting score is somewhat different for nearly all of the scales. In light of these findings, some will be inclined to use the optimal cutting score instead of the recommended score of 30. It should be noted that a score that may be seen as optimal, in the sense of minimizing the sum of the false positives and false negatives, for one sample may not be optimal for another. In other words, the optimal cutting score is a sample statistic, and it cannot be regarded as a fixed parameter. It should also be noted that the cutting score for the separate scales seems to fluctuate around a score of 30. Indeed, at the bottom of Table 8-1 it can be seen that the mean cutting score for all of the scales is 29.8. In light of this finding it seems entirely defensible to recommend that the cutting score of 30 be used for all of the scales.

In addition to the findings concerning the cutting scores for the scales, data have also been provided in relation to the rates of false positives and false negatives. Rates of false positives were computed by determining the proportion of cases, known to be free of the problem being assessed, that were erroneously classified as having a problem when depending on the optimal cutting score. Similarly, the rates of false negatives were computed as the proportion of cases, known to have a problem in the area being assessed, that were erroneously classified as being problem-free when depending on the optimal cutting score. The same procedure was used for the recommended cutting score of 30 for several of the scales.

While the data shown in Table 8–1 provides a reasonable basis for concluding that classification error rates for the CMP scales are small enough to justify their use in clinical and research applications, it is also apparent that much more work is needed in this area. The data that were used for the validation of the CAM and CAF scales are no longer available so it was not possible to report classification error rates for these two scales. Also, formal data are not yet available for the IPR scale but that information should be forthcoming in the near future. In short, there is room for much continued work on several of the CMP scales in terms of obtaining improved estimates of the cutting scores and the classification error rates. The major improvement, it is expected, will come from the conduct of future studies that are based on much larger samples.

SUMMARY

Cutting scores, false positives, and false negatives are important characteristics of assessment devices that are used in clinical practice. Error rates in relation to the classification of clients provide useful guides to the risk one takes in rendering a prothetic diagnosis and the cutting score has several applications in both practice and in research. In the case of the CMP scales, it is convenient that a single score of 30 can be used as an effective cutting score for each of the scales. It was surprising to the author and his colleagues that it turned out that way, and one can only speculate about the reasons for such good fortune. Most likely the high similarity among the scales with respect to the cutting score arose largely because each of the scales has the same general structure and scoring procedures. It must be related also to some underlying consistency that exists among therapists with respect to judgments they make about the presence or absence of clinically significant problems in personal and social functioning among clients—similarity in methods and instructions given to therapists helped too, no doubt.

In concluding this chapter, it should be recognized that some may feel the classification error rates shown in Table 8–1 are rather large. Actually, as compared with many other different types of measurement scales, they are generally quite good. In the final analysis, it is believed that the rates reported in this chapter are actually upper-bound rates. Although it was necessary to use clinician judgments as a criterion, it is well known that therapists also make mistakes in diagnosis. Thus, there are two error rates involved in assessing the performance of the CMP scales; those inherent to the scales and those inherent to the therapists. The data presented here include both in a context that attributes them solely to the scales.

9 Concluding remarks

This field manual was designed to serve as a reference guide for those who may wish to use one or more of the CMP scales in clinical or in scientific applications. It is not complete for the simple reason that additional work is needed with respect to the validation of one or more of the scales. Some of that work is in progress now. If this manual and the CMP scales are widely used and favorably regarded, the manual will be revised and updated as soon as additional findings and information are obtained to justify such a revision. Until such time, it is believed that the information contained in this manual is strong enough to support the recommendation that the CMP scales be used in clinical and research applications in a wide variety of settings and human service organizations. To date, there is an enormous body of research and anecdotal evidence to indicate that the CMP scales will do a credible job of measuring what they were designed to measure and helping clinicians to monitor and evaluate client progress in treatment provided that the scales are properly administered, scored, and interpreted.

LIMITATIONS

No measurement tool is perfect. They all make mistakes. While the goal in applied measurement work should always be one of eliminating such measurement mistakes, that is an impossible task. In the final analysis, we must always live and work with a very great deal of uncertainty. However, an important goal in developing and validating the CMP scales has been one of keeping the risk of error to at least an acceptable minimum. The validity and measurement error characteristics of the CMP scales help to do that, but one can never rely entirely upon the raw psychometric properties of any measurement tool—especially in the applied social and behavioral sciences.

In using the CMP scales, the risk of error is greatly increased by becoming sloppy or careless in administering and scoring the scales and by using them for purposes they were not designed to serve. The limitations and restrictions that were described in the text of this manual were not put there to be taken lightly. They must be adhered to if the scales are to perform in the manner intended and with the quality that is claimed for them. However, the currently available evidence strongly indicates that when the limitations and restrictions discussed in the text are respected and followed, the risk of serious error is kept to a minimum.

There can be no doubt that the greatest threat to the validity of the CMP scales is always going to arise through some form of social desirability or malingering responding. This problem can never be completely eliminated. It can be partially controlled and kept to a minimum, and specific suggestions were given in the text for reducing this source of threat to the effective use of the CMP scales. It should be emphasized again that this problem can be best evaluated by cross-checking the CMP scale scores against all other available information concerning the clients' or respondents' problems in the areas being measured. Without a doubt, the most effective way to minimize this risk is for users of the CMP scales to legitimate and validate their need and right to have the information they seek and to make it evident to the respondent that the information will be used for an important and worthy purpose. For certain, the very best way to maximize the risk of social desirability responding is to provide some overt or subtle clue or hint to indicate that the information will be used to trick, ridicule, embarrass, or otherwise harm or distress the respondent.

A SELF/REPORT DATA CAVEAT

The CMP scales were developed primarily for use by clinicians and researchers who choose to define personal and interpersonal relationship problems in higher order, more abstract, and nonbehavioral terms, but this does not mean that the scales cannot be used effectively by those who adhere to and use behaviorally oriented theory and methods. In order to address some of the measurement needs of nonbehaviorally oriented practitioners and researchers, exclusive reliance was placed upon the use of paper-pencil, self-report measurement techniques.

Self-report measurement devices have a number of inherent weaknesses (Hersen and Barlow, 1976) and these must not be ignored. The plain fact is, however, that one can never completely eliminate the need for using self-report information. Unfortunately, some clinicians and researchers have adopted a zealous stance in claiming that behavioral data and behavioral *observations* are always superior and that self-report data are nearly always suspect. It is a simple truism that an enormous amount of behavioral data that are collected and used in clinical and research applications are themselves self-report measures and that they suffer the same weaknesses as do nonbehavioral self-report data. Moreover, data that are collected in the form of behavioral observations have their own special problems (Hersen and Barlow, 1976), so none escapes.

The intent here is not to beg the question in behalf of the CMP scales, and no attempt is being made to suggest that the problems involved in the use of self-report measures (of any type) should be ignored or minimized. To the contrary, they should be well understood and dealt with, and the discussions provided by Hersen and Barlow (1976) are a good place to start in understanding the limitations involved in using such data. Rather, the intent is to suggest that self-report measures of many forms and types, properly handled, can serve as powerful aids to practitioners in the helping professions who must perform an enormous range of clinical and research functions and tasks.

CONTINUED VALIDATION

Because there are distinct gaps in the currently available knowledge base concerning the performance of the CMP scales, it will be necessary to extend and continue the validation work that has already been conducted and described in previous chapters. At present, there are no formal data to describe the reliability and validity of the IPR scale, and that constitutes the greatest knowledge deficit of the entire package. However, a study is currently in progress (Nurius and Hudson, 1982) that is designed to close that gap. It is fully expected that a final research report will be available within a year of the publication of this manual—and perhaps sooner.

Although several studies have been conducted to partially validate the ISE scale and extant data concerning its reliability appear to be adequate, there is a continued need for better information concerning certain aspects of the validity of the ISE scale. Here again, a special study is in progress to obtain such data (Hudson, Abell, and

Jones, 1982), and a final report of those study findings also should be available within a year of publication of this manual.

The kinds of data that are reported in this manual and the cited studies are very important for those who wish to examine, consider, or use the CMP scales. However, there are other types of information that are also exceedingly important but that are obtained largely through the day-to-day use of the scales in clinical practice. Most of the nonstatistical information about the use and performance of the CMP scales has come from practitioners who have been willing to pass on their observations and experiences in using the scales in their work. More of this type of information is needed, and it is hoped that those who use the scales will feel free to contact the author regarding any special uses, strengths, applications, problems, or weaknesses they feel might help others to make best use of the scales.

To date, the overwhelming majority of the validation research for the CMP scales has been conducted by the author and his colleagues, but it is hoped that others will conduct their own full or partial validations of these instruments. While the evidence reported thus far is presented with the conviction that it will be validated through replication studies conducted by others, such replications have not yet been carried out. The one exception to this statement is the excellent study conducted by Byerly (1979). The intent here is not to pass on to others the burden of demonstrating the validity, reliability, and utility of the scales, but to obtain independent evidence to either support or refute the claims set forth in this manual. Whatever the outcome of such efforts by others, they will provide extremely important findings that practitioners and researchers can use as a basis for judging the quality of data that are obtained from the use of these measurement devices. The author would like very much to have copies of such studies that other researchers would be willing to make available.

INDEPENDENT RESEARCH FINDINGS

Actually, a number of independent studies are beginning to emerge from a variety of uses of the CMP scales. For example, Calhoun (1979) used the ISE scale as a basis for assessing the effectiveness of teaching services provided by a professional fashion model to help adolescent girls reduce self-esteem problems. Although sample sizes were small and the total treatment package consisted of only six training sessions for three different groups, the gains in self-esteem were statistically significant for all three groups. In a complex study of adjustment among homosexual and heterosexual couples, Dailey (1979) used four of the CMP scales and found that they appeared to be sensitive in detecting both small and large group differences. In a rather sophisticated study of the relationship between alcohol abuse and sexual functioning, Holosko (1979) found that the ISS scale was the only one of several measures of sexual dysfunction that appeared to be sensitive enough to produce statistically significant findings in this regard. He reports other findings concerning the use of the ISS with chronic alcoholics that may be of interest to those who work in the area of alcohol abuse. Glisson, Melton, and Roggow (1981) used several of the CMP scales to investigate the effects of separation on personal and interpersonal functioning among couples who experienced involuntary separation when the male partner fulfilled a nuclear submarine sea-duty mission. They found the scales to be reliable and sensitive to family-life changes as a function of separations due to military duty assignments.

While it is too early to make a final judgment, the above studies provide additional evidence to suggest that the CMP scales are construct valid and may be sensitive to experimental effects and survey group differences that other measurement approaches may fail to detect. When other studies are made available over the next several years a great deal of additional information will be obtained concerning the research utilization and limitations of the CMP scales.

THE OTHER LANGUAGE VERSIONS

The CMP scales and this manual were developed primarily for those who work with English-speaking individuals and populations who grew up or who were acculturated in the United States. Quite frankly, very little is currently known about any special problems that may arise in using the nonEnglish versions of the scales. The other language translations of the CMP scales were provided through the courtesy of the translators and on the basis of their convictions that those translations will be useful to practitioners in other countries and to those who work with certain nonEnglish speaking populations in the United States. The judgments of the translators are respected in this regard because they are accomplished professionals who have extensive experience in working with people who speak their language and grew up in different cultures. Even so, it is hoped that practitioners and scientists from other cultures and countries will conduct their own validation studies and provide reports of how the CMP scales best can be used and interpreted in their environments.

SCALE REVISIONS

Over the past few years an effort has been made to resist any revisions of the scales until an adequate body of evidence would suggest or confirm that such revisions were needed. As of this date, there appears to be no need for revision of any of the items on the ISE, IPA, CAM, CAF, or IFR scales. Minor changes have been made in the GCS, IMS, and ISS scales in an effort to improve them, and the factorial validity data presented in Chapter 7 indicates that such changes were needed and justified. Of course, no data are currently available for an assessment of the revised items but future and continued use of the CMP scales will show whether the few changes that have been made constitute improvements or detractions. It is hoped that no further changes will be required for any of the scales and none will be made until new and stronger information demands it.

SUMMARY

This manual has presented and described nine short-form scales that have been shown to be highly reliable and valid measures of a number of personal and interpersonal relationship dysfunctions or problems that are commonly seen in a wide variety of clinical settings. The scales were originally designed to be used in repeated measures designs and applications to provide an improved basis for monitoring and evaluating clinical practice for the purpose of improving the likelihood of a more positive outcome for the client. Thus far, the CMP scales have been used in an enormous number of quite different clinical settings and applications across the United States and in more than 15 foreign countries. Although many persons who are using the scales or who used them in the past have not reported their observations concerning the performance of the CMP scales, the available reports overwhelmingly show that the scales do an outstanding job of helping clinicians to obtain improved estimates of the severity of client problems. Most important, they appear to do an excellent job of helping therapists obtain more reliable and effective assessments of the direction and magnitude of treatment outcomes. Evidence strongly suggests that use of the scales (as well as other similar measurement devices) as described in Chapter 5 has had the effect of helping therapists to significantly improve the quality of their practice for a large proportion of those who adopt the assessment strategy described in this manual.

Although the primary purpose for developing and validating the CMP scales was to use them as diagnostic and assessment devices in clinical practice, they appear to have a few other valuable uses. First, they have been and can be used in a very wide variety of idiographic and nomothetic research applications. Second, they can be used in a variety of training applications in both classroom and practicum settings. In this regard, they appear to be highly useful for the training of both clinicians and researchers. Finally, they have been successfully used as aids in a variety of supervisory and consultative applications.

I clearly do not wish to make overblown claims concerning the performance of the CMP scales. Nonetheless, in concluding this manual it seems appropriate to point out that the number of successful uses of the scales thus far overwhelmingly exceeds the number of failures. Even so, much remains to be done and it is hoped that those who choose to use the scales will feel free to submit their positive and negative findings and experiences as fuel for the work ahead.

appendix A
Self-training exercises

This appendix was prepared as an aid to those who have never used the CMP scales and to those who may not have had any prior experience or training in the use of formal measurement tools in clinical practice. By reading the first five chapters of this manual and completing only a few of the exercises described in this appendix, you should have no difficulty in using any of the CMP scales with your clients.

Before you do these exercises, you should carefully read and think about all of them. Each of them is simple to do and they are designed to help you better understand the use and interpretation of the CMP scales. However, some of the exercises described in this appendix can be very threatening. They will, in some cases, test your mettle with respect to self-disclosure and with respect to the ways you defend yourself and manage your personal affairs. Consequently, some of them could prove harmful or distressing if they are conducted carelessly or irresponsibly. If you feel that completion of any of these exercises could be harmful or distressing to you or to others, do not do them. In short, you must act in a professionally responsible manner in using any of the CMP scales and in conducting these training exercises. If you are uncertain about how to properly complete the exercises or use the scales, seek help and consultation.

EXERCISE 1

Before using any of the CMP scales with clients, you should complete every one of the scales that is appropriate to you and your

FIGURE A–1

Item	GCS	ISE	IMS	ISS	IPA	CAM	CAF	IFR	IPR
1									
2									
3									
4									
5									
6									
7									
8									
9									
10									
11									
12									
13									
14									
15									
16									
17									
18									
19									
20									
21									
22									
23									
24									
25									
Score									

personal circumstances. If that has not been done, do it now. Answer each item as carefully and as accurately as you can. Turn to Chapter 1 and complete the scales now. If you do not want to record your answers on the scales shown in Chapter 1, you may record them on the answer sheet shown below as Figure A–1. When you complete the scales, note the time and see how long it takes you to complete all of the scales. When you've finished, divide the total time by the number of scales you completed to get an idea of how long it takes to complete each one.

EXERCISE 2

Now that you've completed all the CMP scales that apply to you, compute the total score for each one as described in Chapter 2. Time yourself again to see how long it takes you to score each of the scales. At first it should take no more than five to seven minutes, and with practice you will complete it in three minutes—or less. Check your work carefully. Did you have any difficulty following the scoring procedure? Did you remember to reverse-score the proper items on each scale?

EXERCISE 3

Pretend that you omitted items 3, 5, 8, and 9 on both the GCS and ISE scales and then recompute your score. Remember, the scoring procedure is a bit different now. If you have forgotten the procedure, go back to Chapter 2 and read the scoring procedure again.

EXERCISE 4

Compare the GCS score you computed in Exercise 2 with the one you obtained in Exercise 3. They will not likely be identical but they should be very close. They should be within five points of each other. Are they? Do the same comparison with the ISE scale. This exercise will give you additional scoring experience and it should demonstrate quite clearly that a few item omissions will not have a major effect on the total score. If the two GCS scores are very far apart, that likely means you have scored the scales incorrectly. The same holds for all of the scales. Check your work again.

EXERCISE 5

The first four exercises are designed to give you experience in scoring the scales. You should work with the scoring procedures until you are completely comfortable with them and are confident you can use them correctly. The next step is to examine your own scores from an interpretive point of view. Start with the GCS scale. As a trained therapist, you should have a reasonably good idea of just how depressed you are—on a subjective or "gut" level. Does your obtained GCS score match that subjective impression or assessment? If not, how far off is it? Did you get a surprise when you saw your final GCS score or was it pretty much on target? The important point of this exercise is to study all your CMP scale scores in relation to your subjective assessments of your own personal situation. This will help you to better understand how your clients may be feeling when they complete the scales.

EXERCISE 6

This is an exercise in empathy. Imagine that you are a depressed client. Try to see yourself in a real funk, and think of yourself as a person who has a clinically significant depression. You feel lousy and you cry a lot. You believe that no one cares about you and you have a lot of self-pity inside. But, you're definitely not suicidal. Now, complete the GCS in terms of the way you are feeling and score it. Can you develop a better sense of how a client might feel who scores

50 on the GCS? How about 65? How about 21? Try this sort of empathy exercise with several of the other scales. It takes time to do this and it is emotionally taxing, but it is worth it in terms of helping you to understand your clients' sense of distress.

EXERCISE 7

Select one of the scales that you have completed and prepare a chart like the one shown as Figure 5–1. Now, plot your score on the chart and then complete the same scale once a week on the same day as the one on which you completed the first scale. Do this for four weeks and plot your scores on the chart. Study the amount of fluctuation in the scores from week to week. Do your scores appear to be quite stable? Do they tend to rise or decline over the five-week period? If there are any marked changes in the scores, do they change in response to special events that you have experienced? If your scores appear to be very stable over the five-week period, does that represent what you regard to be an accurate assessment of your personal situation? Did your scores tend to drift upward or downward? If so, what does that mean to you? If such a drift did occur, it should not have exceeded five points unless there was real change. If you want to speed up this exercise, you could complete the scales each day for, say, a week and plot your scores daily. This is not recommended when working with clients but it is a useful exercise to do for yourself.

EXERCISE 8

Think back to a time in your life when you were really depressed. Nearly all of us have had such experiences. Recall the feelings that you had at that time and then complete the GCS scale in terms of the way you felt at that time. How large was your score? Was it clinically significant? How well do you think the score matches up to your subjective assessment of how you felt at that time? If you have had some problems in other areas, repeat this exercise by using other CMP scales. How do these retrospective scores help you to better understand or appreciate how your clients might be feeling when they obtain similar scores?

EXERCISE 9

This exercise is for those who are married or who have a partner in an ongoing dyadic relationship. Fill out the IMS scale and ask your partner to do the same thing. Now, fill out the IMS scale again, but this time, fill it out in terms of the way you *think* your partner will fill it out. Ask your partner to do the same thing. Now, compare your partner's IMS score with the one you got when you completed the IMS the way you *thought* your partner would respond. How close are the two scores? Are the two scores within five points of one another? Ten points? More? What do you think that means? In most cases, large differences indicate that one partner is seriously misperceiving the way the other one feels about the relationship. Did you feel threatened by this exercise? If so, what does that mean? How did you deal with it? What do you think this could mean for clients? How candid were you in your own responses? Have your partner compare the two IMS scores: the one you got and the one your partner got when he or she completed the IMS the way he or she *thought* you would complete it. Does this exercise suggest that one or both of you are very confident about your relationship with each other or that one or both of you may be misperceiving the other—in what direction?

EXERCISE 10

In this exercise, use the responses you obtained from Exercise 9. Think about the IMS scores that each of you obtained when you completed the scale the way you felt about the relationship. Both of you are likely to feel that the IMS score tells you nothing about your relationship that you did not already know. Now, exchange scales and examine how each of you responded to the items. Be careful. Each of you may be flattered by some of the responses but you may feel threatened, offended, embarrassed, or angered by others. Do not do this exercise unless you are prepared to deal with the answers you get. Each of you is likely to learn something about how the other feels that you were not aware of before. Did that happen? For one or both of you—or neither? Discuss your differences and try to understand what they mean.

This exercise is designed to show you how the CMP scales can sometimes be used as a communication device. However, if you ever use this strategy in practice, take good precautions so that you do not violate clients' rights to confidentiality and privacy. Also be mindful of the fact that this strategy could mislead you or the clients if one or both of them are unwilling to provide candid disclosures. Also, this is obviously not a good procedure to use when one or both partners demonstrate a pattern of searching for excuses or opportunities to aggress against the other partner. For such couples, this sort of exercise could have the decided disadvantage of giving one partner a lot of ammunition to use against the other partner.

EXERCISE 11

In spite of the fact that it is very easy to administer and score any of the CMP scales, it is surprising to learn that many therapists are initially very uncomfortable when they ask clients to complete the scales for the first time. If you feel that you might experience this type of discomfort, you should role-play a client interview either alone or with someone else at least once or twice before using the scales in practice. If you can thereby eliminate or reduce your anxiety in this regard, less of it will be communicated to the client. The client will therefore obtain from you a much better understanding of why you want the scales completed and how they will be used in treatment. Most important, the client will be much more willing to cooperate. If you are in training, it is good to conduct such role-play exercises in a training group and then obtain feedback from your colleagues.

With a little imagination you can think of many other exercises that you can carry out to help you gain experience and confidence in using the CMP scales in clinical practice. By all means, create your own exercises and use them. However, you do not have to spend an inordinate amount of time to learn how to use the scales successfully. The exercises described above should be quite adequate to enable you to begin using the scales rather quickly with your clients. Bear in mind, however, that few people can ever read a manual, such as this, and remember, after one reading, all that it contains. Read it thoroughly five or six times (especially Chapters 1 to 5), and you should then find that you will have a powerful and flexible set of tools that you can use to help increase the likelihood of more positive outcomes for your clients.

appendix B
A general screening device

In Chapter 8 the Psycho-Social Screening Package or PSSP was introduced, and exhibited as Figure 7–1, in the context of examining the construct validity of the CMP scales. The major purpose for developing the PSSP was to have a method of efficiently collecting information from clients about a wide range of personal and interpersonal relationship disorders in order to then correlate the PSSP items with the CMP scales.

After working with the PSSP in several different studies, it began to appear that it could be used as a crude screening device to help clinicians obtain a quick and effective overview of client problems. It can be used as an intake screening device and it can be used during the initial and final stages of treatment as one way of assessing the general status and well-being of the client. If one chooses to use the PSSP as a screening tool, it is important to realize that it does not have the psychometric power of the CMP scales; it has a very different purpose and structure; it is scored and interpreted very differently; and it should not be used as a *primary* assessment tool. However, if due regard is given to the limitations of the PSSP, it can serve as another effective aid to assessment and evaluation.

THE PSSP STRUCTURE AND PURPOSE

The PSSP is quite different in structure and purpose from the CMP scales. Each of the CMP scales contains 25 items that are used to measure *one* construct or problem. The PSSP contains 20 items that are used to measure 20 different problems. The PSSP is therefore designed to provide the therapist with a quick and efficient look at many potential problems for a single client or respondent. Each of the 20 items comprising the PSSP is structured as a category partition scale and each has a score range from one to five.

The major purpose of the PSSP is to obtain a rapid overall assessment of client problems and to identify how many specific problems may need attention; which ones are prepotent; and how serious is each of them. The 20 problems represented by the PSSP do not reflect the total range of problems the therapist must check out or investigate and the PSSP should not be used in a manner that could encourage therapists to ignore other potential problem areas. It does, however, span the range of problems that are commonly seen among nonpsychotic clients in outpatient settings.

Nine of the items on the PSSP correspond to the problem constructs that are measured by the CMP scales. If one or more of those nine PSSP items indicate that the client may have a problem, the therapist should follow this up by having the client complete the appropriate CMP scales. The PSSP items will provide a rough index of the seriousness of a specific problem, but they do not have the same precision that one can obtain from the CMP scales. If the PSSP indicates the client has one or more problems in areas not measured by the CMP scales, the therapist should conduct a more thorough assessment of those problems to obtain a more precise understanding of them.

SCORING AND INTERPRETING THE PSSP

Because the PSSP measures 20 different problems, the items are not scored and interpreted in the same manner as the CMP scales. First, there are no item reversals on the PSSP. Second, the PSSP items are not summed and interpreted in the same way as the CMP scales. Third, there are several different ways to "score" the PSSP, and, finally, it is important to prepare a *profile chart* for the PSSP.

The PSSP item scores

As indicated above, each of the PSSP items is designed to measure a different problem. Thus, unlike the CMP scales in which one is discouraged from giving much attention to specific item responses, each of the PSSP items is interpreted separately. The general findings obtained thus far in using the PSSP in several studies indicates that a score of 4 or 5 on any PSSP item constitutes fairly strong evidence of a clinically significant problem in the area that is represented by that item. Similarly, a score of 1 or 2 usually means the client does not have a clinically significant problem in the area that is represented by the item. If a client scores a PSSP item as three, that usually means there is a good possibility of a clinically significant problem, but that should be checked out very carefully by obtaining more information from other sources.

Those who may choose to use the PSSP as a screening or initial assessment device must be acutely aware of the fact that the PSSP items are not nearly as reliable as the CMP scale scores. There is a great deal of measurement error inherent in most types of single-item measurement tools and the PSSP items are no exception. There is no doubt about their validity, because the content of each item is so sharply focused on a specific problem or target area. In other words, the PSSP item responses can be used as a preliminary guide in assessing whether a client has a problem in one or more of 20 different areas and they can provide a rough idea of how serious those problems might be. However, the PSSP items should never be used as the primary basis for forming a diagnosis or a treatment plan.

PSSP summary scores

It is entirely possible to compute a variety of overall summary scores by using the PSSP item responses. Some may be useful while others are strongly recommended against using. If certain types of summary scores are computed and used in either clinical or research settings or applications, they can seriously mislead the therapist or the researcher. To reduce the risk of that happening, it is important to examine carefully the types of summary scores that should not be used and those which will be of some use.

The PSSP total score. It is entirely possible to compute an overall total score for the PSSP by summing all the items that are scored from 1 to 5 and dividing that sum by the number of such items. Such a total score would be computed as $T = \Sigma X/N$ and it would have a possible range from 1.0 to 5.0. This type of total score should *never* be computed for the PSSP because it can be severely misleading. It is important to understand why.

Suppose a client obtains a score of 1 on 19 of the PSSP items but scores one item as 5. If that happens, the total score will be 1.2 even though the client has a serious problem in one of the areas covered by the PSSP. In fact, a client could actually score five of the items as 5 and still obtain a total score of only 2.0 provided that the other 15 items were all scored as 1. From these examples, one can see that *such a total score can be dangerously misleading.* It is best not to use it.

The PSSP subscores. The PSSP has a logical dimensional structure that one might use to compute subscores. The first 12 items of the PSSP all relate to some form of personal distress and the last eight items all relate to some form of interpersonal relationship disorder. If the above scoring procedure is applied only to the first 12 items, one could apparently obtain something that may be called a "personal distress" or PD score. Also, if one applies the same scoring procedure to the last eight items, one could apparently obtain something that may be called an "interpersonal distress" or ID score. These types of PD and ID scores suffer the same problem as does the total score, and *they should never be used.*

The PSSP problem index. The best and safest way to obtain an overall index or score for the PSSP is to compute what is known as the "problem index" or PI score. That is done by summing *only* those items that are scored as three, four, or five and then dividing that sum by the number of such items. The PI score has a possible range from 3.0 to 5.0 provided that at least one of the PSSP items was scored from three to five. If none of the PSSP items is scored from three to five, then and only then is the PI score made equal to the total score described above. Thus, in the final analysis, the PI score does have a possible range from 1.0 to 5.0.

The PPI and IPI scores. It is entirely possible to safely take advantage of the dimensional structure of the PSSP by applying the PI scoring procedure separately to the two sets of items on the PSSP. That is, if the PI scoring procedure is applied only to the first 12 items on the PSSP, one can obtain a "personal problem index" or PPI score that may be useful in some cases. Similarly, if the PI scoring procedure is applied to the last eight items of the PSSP, one can obtain an "interpersonal problem index" or IPI score. Both the PPI and IPI scores will range from 3.0 to 5.0 for any client who has one or more problems in these two domains. If the client does not have a clinically significant problem, the PI, PPI, and IPI scores will all range from 1.0 to 2.0, so the overall range for all three of these scores will be from 1.0 to 5.0.

To illustrate the proper scoring of the PSSP, consider the case of Mr. L. and the responses he gave as shown in Figure B–1. The PI score for Mr. L. was computed by summing items 1, 3, 7, and 8 and dividing by 4 to get PI = 4.0. Because all four of the items scored three or more are among the first 12, the PPI score for Mr. L. will also

FIGURE B-1

PSYCHO-SOCIAL SCREENING PACKAGE (PSSP)

Mr. L.
Case # 45L67

This questionnaire provides you a means of reporting problems that you may be having as an individual or in your relationships with your family and others whom you know or work with. It is not a test, so there are no right or wrong answers. Answer each item as carefully and as accurately as you can by placing a number beside each one as follows:

1 Rarely or none of the time
2 A little of the time
3 Some of the time
4 Good part of the time
5 Most or all of the time
6 This item does not apply to my situation

Please begin.

Item	Statement	Score
(1.)	I feel depressed	4
2.	I have a low sense of self-esteem	2
(3.)	I feel unhappy	5
4.	I feel afraid	1
5.	I feel anxious	2
6.	I feel nervous	2
(7.)	I have disturbing thoughts or fantasies	4
(8.)	I have problems with my own anger or rage	3
9.	I have nightmares	2
10.	I have a problem with my own sense of identity	2
11.	I have problems with my personal sex life	2
12.	The quality of my work is poor	1
13.	I have problems in my relationships with friends	2
14.	I have problems in my relationships with people I work with	1
15.	There are problems in our family relationships	2
16.	I have problems with one or more of my children	1
17.	My partner and I have problems in our relationship	1
18.	My partner and I have problems with our sexual relationship	1
19.	My mother and I have problems in our relationship	1
20.	My father and I have problems in our relationship	1

be 4.0. None of the last eight items was scored three or more, so Mr. L.'s IPI score is computed as the simple average of the last eight items on the PSSP to obtain IPI = 1.25. Now consider the scores obtained from Mrs. C. as shown in Figure B-2. Her overall PI score is 3.71; her PPI score is 3.8; and her IPI score is 3.5.

Again, it is very important that one does not compute any form of

FIGURE B-2

PSYCHO-SOCIAL SCREENING PACKAGE (PSSP)

Mrs. C.
Case # 45L67

This questionnaire provides you a means of reporting problems that you may be having as an individual or in your relationships with your family and others whom you know or work with. It is not a test, so there are no right or wrong answers. Answer each item as carefully and as accurately as you can by placing a number beside each one as follows:

1 Rarely or none of the time
2 A little of the time
3 Some of the time
4 Good part of the time
5 Most or all of the time
6 This item does not apply to my situation

Please begin.

	Item	Score
1.	I feel depressed	1
2.	I have a low sense of self-esteem	2
3.	I feel unhappy	3
4.	I feel afraid	5
5.	I feel anxious	4
6.	I feel nervous	4
7.	I have disturbing thoughts or fantasies	3
8.	I have problems with my own anger or rage	1
9.	I have nightmares	2
10.	I have a problem with my own sense of identity	1
11.	I have problems with my personal sex life	1
12.	The quality of my work is poor	1
13.	I have problems in my relationships with friends	3
14.	I have problems in my relationships with people I work with	4
15.	There are problems in our family relationships	2
16.	I have problems with one or more of my children	1
17.	My partner and I have problems in our relationship	1
18.	My partner and I have problems with our sexual relationship	1
19.	My mother and I have problems in our relationship	2
20.	My father and I have problems in our relationship	6

PSSP summary score that fails to distinguish between items that are scored above two and those that are scored as one or two. The major risk of doing that is to provide a summary score that could make it appear the client has no clinically significant problem when such a conclusion could be definitely false. The major risk is to markedly increase the probability of a false negative score.

Even though the PI, PPI, and IPI scores can be used without increasing the risk of obtaining a false negative, they are nonetheless difficult to interpret from a substantive or construct point of view. Because they mix up several different problems (or are capable of doing so), their only utility is to provide a very general index of client problem level. Only a close look at the specific items will provide information about which specific areas the client is having problems with. A much more effective way to use the PSSP is to prepare a profile chart as described below.

THE PSSP PROFILE CHART

When a client completes the PSSP, the therapist has a great deal of information that spans 20 different problem areas. It is sometimes difficult to absorb and integrate that much information with ease and that task can often be aided considerably by use of some type of

FIGURE B-3

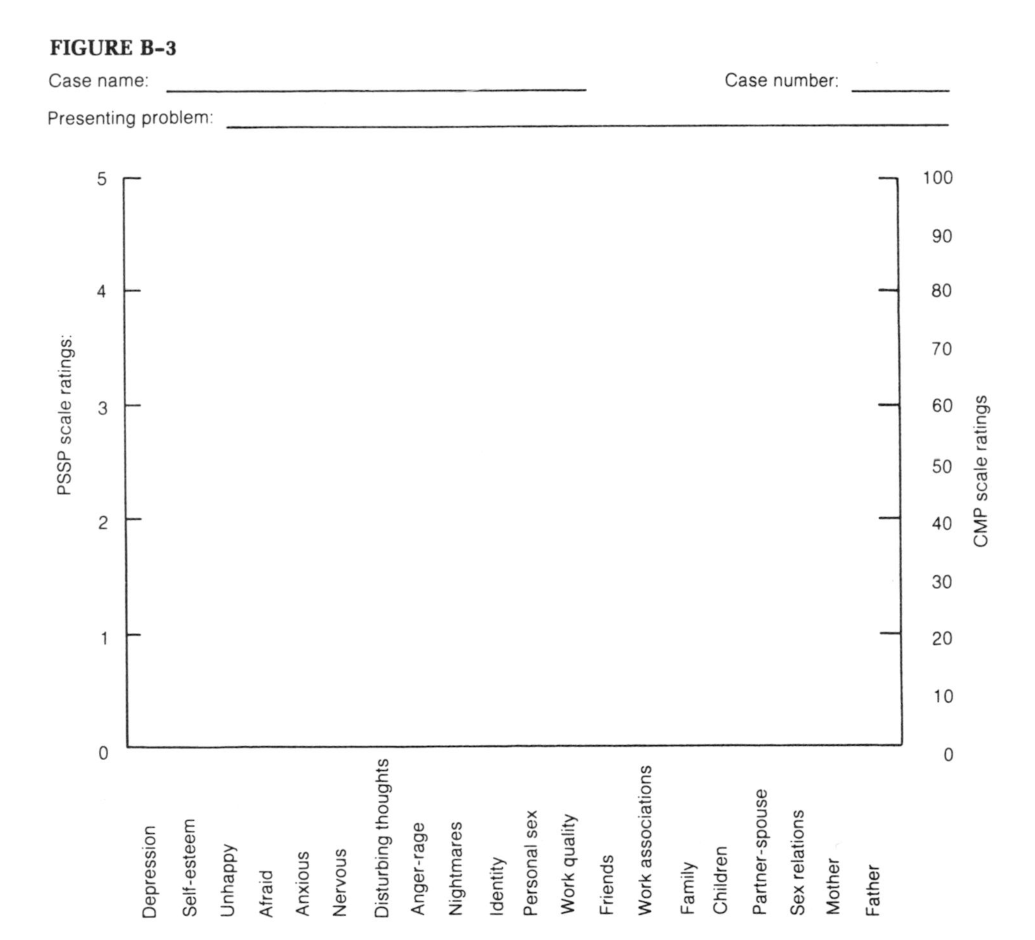

visual aid. Moreover, the overall configuration that is presented by such a device enables the therapist to examine and think about the total profile in a way that is quite different from an inspection of the client's problems one by one.

In order to aid the process of examining and presenting the overall PSSP profile, a special chart has been prepared as shown in Figure B–3. Once the client has completed the PSSP, the therapist can complete the profile chart by shading or crosshatching the column for each item up to the score level indicated by the client's response to the items. As an illustration, the profile chart for the case of Mrs. C. is shown as Figure B–4. In order to facilitate the cross-checking of the PSSP item responses, a separate CMP scale has been provided in the right margin of the profile chart so that therapists can easily record the CMP scale scores as well.

FIGURE B–4

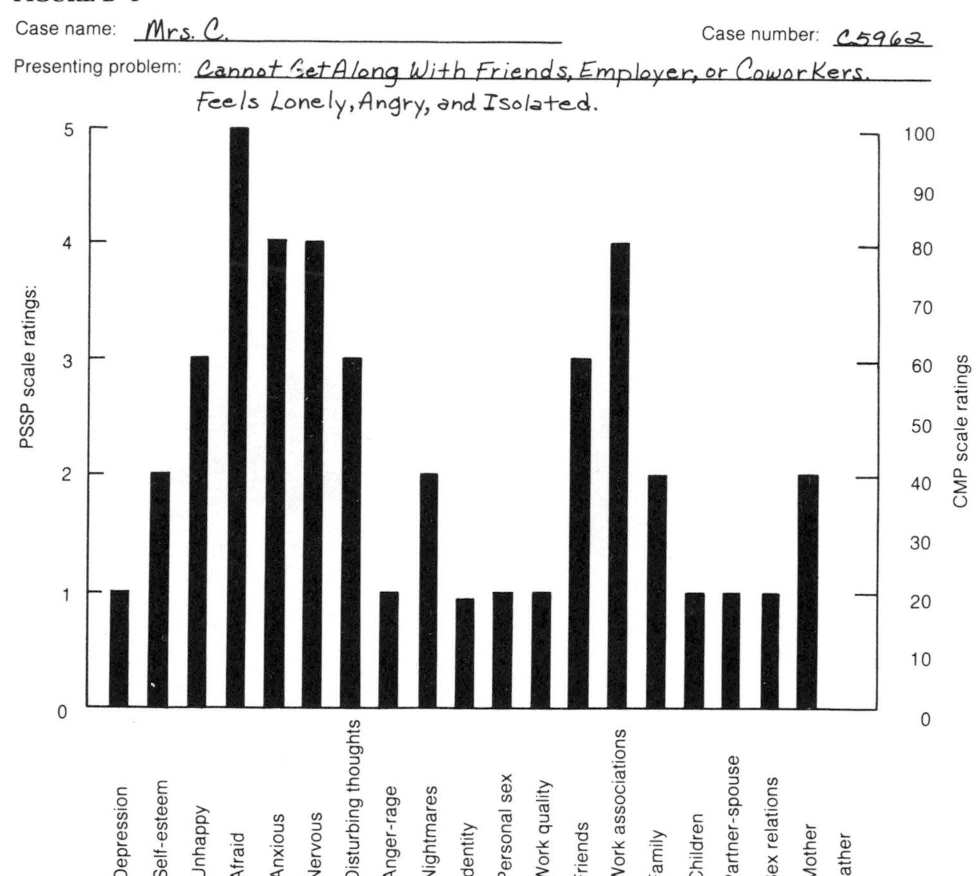

As with the time-series CMP charts described in Chapter 5, the PSSP profile chart should be completed for each client who completes the PSSP and the profile chart should be dated and placed in the case record. The profile chart can be used quickly to remind the therapist of the overall intake assessment for a particular client and it can be used to facilitate communication when working with one's supervisor or a consultant. For short term intervention plans, the PSSP can be administered at the beginning and end of therapy to obtain a type of pre- and posttreatment profile summary of the case and its outcome. For cases that require long periods of treatment, several administrations of the PSSP and their associated profile charts can provide a rough picture of overall changes in the client's problem profile over the course of treatment.

There has not been any research to date that will show whether and to what extent the overall profile configuration for the PSSP provides useful information regarding prognosis over and above that which is obtained from separate problem scores. That body of research will probably grow out of the direct observations made by clinicians after obtaining considerable experience in examining a wide range of different types of profiles. However, the few therapists who have used the PSSP and the profile chart have reported that it helps them to obtain a better idea of what type of case they are dealing with and it helps them to pin down which one or more problems to focus on in formulating an initial treatment plan. Most important, it provides immediate information to help decide which of the CMP scales to select for use in monitoring and evaluating the client's progress over time.

As cautioned earlier, the PSSP and its profile chart are not presented as a substitute for the CMP scales and it should not be used as such. But, if used judiciously and with due regard for its inherent limitations, it may serve as an additional aid to facilitate problem description, diagnosis, case management, treatment planning, and an evaluation of treatment outcome.

References

Bartosh, J. C. *Non-psychotic depression as a function of individual distress and interpersonal relationship dysfunction.* Unpublished master's research project, University of Hawaii School of Social Work, 1977.

Beck, A. T.; Ward, C. H.; Mendelson, M.; Mock, J.; & Erbaugh, J. An inventory for measuring depression. *Archives of General Psychiatry,* 1961, *4*, 561–571.

Byerly, F. C. Comparison between inpatients, outpatients, and normals on three self-report depression inventories. Unpublished doctoral dissertation. Kalamazoo: Western Michigan University Department of Education, 1979.

Calhoun, N. Social awareness group for girls with poor self-concept. *School Social Work Journal,* 1979, *3* (2).

Campbell, D. T. & Fiske, D. W. Convergent and discriminant validation by the multitrait-multimethod matrix. *Psychological Bulletin,* 1959, *56*.

Cattell, R. B. The scree test for the number of factors. *Multivariate Behavioral Research,* 1966, *1* (2).

Cattell, R. B., and Dickman, K. A dynamic model of physical influence demonstrating the necessity of oblique simple structure. *Psychological Bulletin,* 1962, 59.

Cheung, P. P. L., & Hudson, W. W. Assessment of marital discord in social work practice: A revalidation of the index of marital satisfaction. *Journal of Social Service Research,* 1981, *4* (5).

Cronbach, L. J. Coefficient alpha and the internal structure of tests. *Psychometrika,* 1951, *16* (3).

Dailey, D. M. Adjustment of heterosexual and homosexual couples in pairing relationships: An exploratory study. *The Journal of Sex Research,* 1979, *15* (2).

Downie, N. M., & Heath, R. W. *Basic statistical methods.* New York: Harper & Row, 1967.

Eysenck, H. The effects of psychotherapy: An evaluation. *Journal of Consulting and Clinical Psychology,* 1952, *16*, 319–324.

Fischer, J. Is casework effective? A review. *Social Work,* 1973, *18*, (1).

Fischer, J. *The effectiveness of social casework.* Springfield, Ill.: Charles C Thomas, 1976.

Fischer, J. *Effective casework practice: An eclectic approach.* New York: McGraw-Hill, 1978.

Fischer, J., & Bloom, M. *Evaluating practice: Guidelines for the helping professional.* New York: Prentice-Hall, 1982 (in press).

Gambrill, E. D. *Behavior modification: Handbook of assessment, intervention, and evaluation.* San Francisco: Jossey-Bass, 1977.

Garfield, S. L., & Bergin, A. E. *Handbook of psychotherapy and behavior change.* New York: John Wiley & Sons, 1978.

Giuli, C. A., & Hudson, W. W. Assessing parent-child relationship disorders in clinical practice: The child's point of view. *Journal of Social Service Research,* 1977, *1* (1).

Glisson, C. A.; Melton, S. C.; & Roggow, L. The effect of separation on marital satisfaction, depression, and self-esteem. *Journal of Social Service Research,* 1981, *4* (1).

Gottman, J. M., & Leiblum, S. *How to do psychotherapy and how to evaluate it: A manual for beginners.* New York: Holt, Rinehart & Winston, 1974.

Grinnell, R. M., Jr. *Social work research and evaluation.* Itasca, Ill.: F. E. Peacock, 1981.

Helmstadter, G. C. *Principles of psychological measurement.* New York: Appleton-Century-Crofts, 1964.

Hersen, M., & Barlow, D. H. *Single-case experimental designs: Strategies for studying behavior change.* New York: Pergamon Press, 1976.

Holosko, M. J. *Sexual dysfunctioning of males undergoing treatment for alcoholism.* Unpublished doctoral dissertation, University of Pittsburgh School of Social Work, 1979.

Hontanosas, D.; Cruz, R.; Kaneshiro, K.; and Sanchez, J. *A descriptive study of spouse abuse at the University of Hawaii at Manoa.* Unpublished master's research project, University of Hawaii School of Social Work, 1979.

Hudson, W. W. *A measurement package for clinical workers.* Paper presented at the Council on Social Work Education Annual Program Meeting, Phoenix, February 1977. Revised and produced by the University of Hawaii School of Social Work, 1978a. Revised and published in *The Journal of Applied Behavioral Science,* 1981a (in press).

Hudson, W. W. Research training in professional social work education. *Social Service Review,* 1978b, *52* (1).

Hudson, W. W. Development and use of indexes and scales. In R. M. Grinnell, Jr. (Ed.), *Social work research and evaluation.* Itasca, Ill.: F. E. Peacock, 1981b.

Hudson, W. W.; Abell, N., & Jones, B. *A revalidation of the index of self-esteem.* Unpublished manuscript. Tallahassee: Florida State University School of Social Work, 1982.

Hudson, W. W.; Acklin, J. D.; & Bartosh, J. C. Assessing discord in family relationships. *Social Work Research & Abstracts,* 1980, *16* (3).

Hudson, W. W., & Glisson, D. H. Assessment of marital discord in social work practice. *Social Service Review,* 1976, *50* (2).

Hudson, W. W.; Hamada, R.; Keech, R.; & Harlan, J. *A comparison and revalidation of three measures of depression.* Unpublished manuscript. Tallahassee: Florida State University School of Social Work, 1980.

Hudson, W. W.; Harrison, D. F.; & Crosscup, P. C. A short-form scale to measure sexual discord in dyadic relationships. *The Journal of Sex Research,* 1981, *17* (2).

Hudson, W. W.; Harrison, D. F.; & Maxwell, S. *The abuse of female intelligensia.* Unpublished manuscript. Tallahassee: Florida State University School of Social Work, 1982.

Hudson, W. W.; Hess, L.; & Matayoshi, P. *Affect disorders and academic achieve-*

ment. Report prepared for the staff of Midpacific Institute. University of Hawaii School of Social Work, 1979.

Hudson, W. W., & McIntosh, S. R. The assessment of spouse abuse: Two quantifiable dimensions. *Journal of Marriage and the Family,* 1981, 43 (4).

Hudson, W. W., & Murphy, G. J. The non-linear relationship between marital satisfaction and stages of the family life cycle: An artifact of Type I errors? *Journal of Marriage and the Family,* 1980, 42 (2).

Hudson, W. W.; Murphy, G. J.; & Nurius, P.S. *A short-form scale to measure liberal vs. conservative orientations toward human sexual expression.* Unpublished manuscript. Tallahassee: Florida State University School of Social Work, 1981.

Hudson, W. W.; & Nurius, P. S. *Sexual activity and preference: Six quantifiable dimensions.* Unpublished manuscript. Tallahassee: Florida State University School of Social Work, 1981.

Hudson, W. W., and Proctor, E. K. *Assessment of depressive affect in clinical practice.* Unpublished manuscript, University of Hawaii School of Social Work, 1976a.

Hudson, W. W.; and Proctor, E. K. *A short-form scale for measuring self-esteem.* Unpublished manuscript, University of Hawaii School of Social Work, 1976b.

Hudson, W. W., & Proctor, E. K. Assessment of depressive affect in clinical practice. *Journal of Consulting and Clinical Psychology,* 1977, *45* (6).

Hudson, W. W.; Wung, B.; and Borges, M. Parent-child relationship disorders: The parent's point of view. *Journal of Social Service Research,* 1980, *3* (3).

Kratochwill, T. R. *Single subject research: Strategies for evaluating change.* New York: Academic Press, 1978.

Lubin, B. *Manual for the depression adjective check lists.* San Diego: Educational and Industrial Testing Service, 1967.

McIntosh, S. R. *Validation of scales to be used in Research on spouse abuse.* Unpublished master's thesis, University of Hawaii Department of Psychology, 1979.

Mullen, E. J.; Dumpson, J. R.; & Associates. *Evaluation of social intervention.* San Francisco: Jossey-Bass, 1972.

Murphy, G. J. *The family in later life: A cross-ethnic study in marital and sexual satisfaction.* Unpublished doctoral dissertation. New Orleans: Tulane University, 1978.

Murphy, G. J.; Hudson, W. W.; and Cheung, P. P. L. Marital and sexual discord among older couples. *Social Work Research & Abstracts,* 1980, *16* (1).

Nunnally, J. C. *Psychometric theory.* New York: McGraw-Hill, 1978.

Nurius, P. S. Mental health implications of sexual orientation. *The Journal of Sex Research,* 1982 (in press).

Nurius, P. S., & Hudson, W. W. *The assessment of peer discord in clinical practice.* Unpublished manuscript. Tallahassee: Florida State University School of Social Work, 1982.

Overall, J. E., and Klett, C. J. *Applied multivariate analysis.* New York: McGraw-Hill, 1972.

Platt, J. R. Strong inference. *Science,* 1963, *146,* 347–353.

Rosenblatt, A., & Waldfogel, D. *Handbook of clinical social work.* San Francisco: Jossey-Bass, 1982 (in press).

Sidman, M. *Tactics of scientific research.* New York: Basic Books, 1960.

Stevens, S. S. Ratio scales of opinion. In D. K. Whitla (Ed.), *Handbook of measurement and assessment in behavioral science.* Reading, Mass.: Addison-Wesley, 1968.

Thurstone, L. L. *Multiple factor analysis.* Chicago: University of Chicago Press, 1947.

Wood, K. M. Casework effectiveness: A new look at the research evidence. *Social Work,* 1978, *23.*

Zastrow, C. *The practice of social work.* Homewood, Ill.: The Dorsey Press, 1981.

Zung, W. W. K. A self-rating depression scale. *Archives of General Psychiatry,* 1965, *12.*

Index

This book has been set VIP, in 10 and 9 point Melior, leaded 2 points. Chapter numbers are 36 point Quadrata and chapter titles are 16 point Quadrata. The size of the type page is 26 by 46 picas.